Wildflowers
OF THE PINE BARRENS OF NEW JERSEY

BY HOWARD P. BOYD

Plexus Publishing, Inc.
Medford, NJ

First printing, March 2001

Wildflowers of the Pine Barrens of New Jersey

Copyright © 2001 by Howard P. Boyd

Published by:
Plexus Publishing, Inc.
143 Old Marlton Pike
Medford, New Jersey 08055
U.S.A.

All rights reserved. No part of this book may be reproduced in any form or by any electronic or mechanical means, including information storage and retrieval systems, without permission in writing from the publisher, except by a reviewer, who may quote brief passages in a review.

Printed in Hong Kong

Library of Congress Cataloging-in-Publication Data

Boyd, Howard P.
   Wildflowers of the Pine Barrens of New Jersey / by Howard P. Boyd.
     p. cm.
   Includes bibliographical references (p.).
   ISBN 0-937548-45-6
   1. Wild flowers—New Jersey—Pine Barrens—Identification.
   2. Wild flowers—New Jersey—Pine Barrens—Pictorial works.
   I. Title.

QK175 .B69 2000
582.13'09749—dc21

00-058916

Publisher: Thomas H. Hogan, Sr.
Editor-in-Chief: John B. Bryans
Managing Editor: Janet M. Spavlik
Production Manager: M. Heide Dengler
Copy Editors: Lauree Padgett
              Michelle Sutton-Kerchner
Cover and Book Designer: Erica Pannella
Sales Manager: Pat Palatucci

TO DORIS,

for her support, her companionship, and her love

CONTENTS

Preface ix

Flora of the New Jersey Pine Barrens 1

Anatomy of a Flower 7

Glossary 9

Species Descriptions 15

Literature Cited and Other References 157

Indexes
    Common names 159
    Scientific names 161

PREFACE

When I first broached the idea of this book to Tom Hogan, Sr., president of Plexus Publishing, he reminded me that I'd said in my *Odyssey* book that it would be "my final effort to develop greater awareness of the wonders of the pine barrens and the need to preserve them." He was right. At that time I had meant it. I had no intention of misleading anyone, but one can change one's mind. So I apologize as I offer another effort with the same purpose of stimulating public awareness of the beauties of the pine barrens and the need to preserve them.

If I have to have a rationale for this present effort, it is that over many years I have accumulated thousands of slides on pine barrens subjects, especially wild flowers, which should be put to better and more widespread use than just being shown in a few slide presentations or being stored in file boxes at home. Put very simply, I have the slides and I'd like to share them so that others will have a better appreciation of the beauties of the pine barrens. The result is this small book of 150 descriptions, supported by 130 natural color photographs of the most common and most likely-to-be-seen wild flowers in the pine barrens of New Jersey. This is only a small representation of the approximately 800 species of flowering plants in the pines. Included also are a few unusual "finds" of wild flowers, mainly among the orchids, some of which are not generally recognized as pine barrens species. However, these species exist in isolated locations within the pinelands and help make the pine barrens such a unique place that it simply must be protected and preserved for future generations.

Please note that the book's emphasis is on wild flowers, meaning mainly herbs, plants that grow up out of the ground each spring and die down again each fall without developing any woody tissue. Thus, the flowers of only one tree, swamp magnolia, and a few low shrubs and sub-shrubs, mostly members of the heath family, are included. Not included are any grasses or grass-like herbs such as sedges and rushes because, in the popular sense, these are not generally thought of as wild flowers, even though they are, in fact, flowering and seed-producing herbs. Note also that, with very few exceptions, such as bird's-foot violet and butterfly-weed, emphasis is on native and most characteristic wild flowers rather than on introductions into the pine barrens that are largely the result of human disturbances. Readers will also note the dominance of three families of flowering plants in the pine barrens: the orchids, the heaths, and the composites, for, in spite of the normally severe abiotic conditions in the pines, such as nutrient-poor, acid, sandy soils, acid waters, and frequent fires, members of these families do well in our pinelands.

Faced with many choices concerning the order in which these subjects should be presented—choices such as alphabetical, by color, by habitats, or phylogenetically—I opted, finally, for the seasonal or chronological sequence in the hope that this will prove most useful for the majority of users of this book. So, we'll start in March and end in October and, in between, we'll try to keep the presentation sequence as close to the blossoming times of these flowers as will be practical, given the necessary parameters that must be followed by the publisher. An exception is that some similar species in the same genus, such as pipeworts, ladies' tresses, sundews, milkweeds, asters, bonesets, and goldenrods, are grouped together following the first one of the group to blossom, regardless of their later blooming sequence, so these can be more easily compared and identified.

This book is not intended to be a technical, scientific, botanical manual. Instead, it is intended to assist anyone, even the most uninitiated, to be able to identify the wild flowers they are most likely to see in the pine barrens of New Jersey. Very purposely, species descriptions have been written with this in mind. Instead of highly technical, scientific, botanical language and terminology, these descriptions have been written in common, everyday language so that each and every individual should be able to follow each description, compare it with the related photograph, and then compare the two with the specimen in the field and identify it—all without the need to interpret technical, botanical terms. In spite of this, every effort has been made to write descriptions that are scientifically accurate, and the scientific nomenclature used in this book follows Gleason and Cronquist, 1991, *Manual of Vascular Plants of Northeastern United States and Adjacent Canada*, second edition, published by the New York Botanical Garden, Bronx, New York.

Some may feel I am being presumptuous to author a book describing and illustrating wild flowers in the pine barrens when there are others who are more knowledgeable concerning the subject matter and still others who may be better wild flower photographers. I wish to assure everyone that this book is in no way an attempt to outdo or compete with anyone. Rather, the intent of this book is to continue to spread the word about the beauties of the pine barrens and why critical pine barren habitats must be preserved so that these and other species will survive well into the future.

While on the subject of preservation, please allow me to urge your support of organizations that are dedicated to preservation of the pine barrens. Chief among these is the Pinelands Preservation Alliance whose principal role is to serve as a watchdog organization over actions by the New Jersey State Pinelands Commission and to encourage the conservation and preservation of pine barren habitats, including their flora and fauna. Other organizations worthy of support include the New Jersey Chapter of the Nature Conservancy and the New Jersey Conservation Foundation, both of which have made, and continue to make, major contributions and commitments toward conservation and preservation of critical pine barren habitats.

While most of the photographs are my own, I wish to express grateful appreciation to Mrs. May Borton, daughter of Dr. David H. ("Bart") Ulmer, for her kind permission to use three of her father's fine slides which are so identified. Also, my dear wife, Doris, provided a number of photographs of subjects of which she had taken considerably better shots than I. These also are identified. For these, I am deeply appreciative, as well as for her companionship on the many exploratory and photographic forays we have taken into the pine barrens together over the years, in spite of her dislike of ticks. In addition, she read the entire text and made numerous suggestions for its improvement. I also want to acknowledge the fellowship and shared knowledge of many other naturalists including Karl Anderson, Joe Arsenault, Gerard Bailey, Mike Baker, Ted Gordon, Mike Hogan, Don Kirchhoffer, Lois Morris, Bob Moyer, Augie Sexauer, and Ellen and Ralph Wilen, all of whom, at various times, have gone with me or I with them, on many pleasant exploratory field trips into the pine barrens looking for and identifying wild flowers. Several of the photographs in this book are the direct result of leads given to me by Ralph Wilen who passed away very suddenly during the final stages of this book's publication.

I want to express special thanks and appreciation to Heide Dengler of Plexus Publishing for overseeing the production of this book, and for her persistent efforts to find a printing house that would produce it with the color photographs on the pages facing the descriptions, at a cost that would not be prohibitive.

Finally, I wish to express appreciation to Tom Hogan, Sr., John Bryans, Janet Spavlik, Erica Pannella, Michelle Sutton-Kerchner, Lauree Padgett, Sandy Sutton, Pat Palatucci, and the rest of the Plexus staff, for their invaluable cooperation, assistance, and support throughout the entire process of producing this book.

FLORA OF THE NEW JERSEY PINE BARRENS

The pine barrens of New Jersey is noted worldwide for its unique flora. A distinctive feature of pinelands plants is the number of both northern and southern forms that co-exist here. When the glaciers of the Ice Age moved south from the Arctic, the tons of ice scoured the surface of the Earth, bringing down gravel, rocks, and considerable northern vegetation. Eventually, some 12,000 to l0,000 years ago, the glaciers, including the great Wisconsin Ice that reached northern New Jersey, melted away, and it is believed that the present pine barrens flora developed only after that retreat.

When the glaciers did retreat, the resulting melt flowed south, down over the sandy Atlantic Coastal Plain of southern New Jersey, depositing along the way some of the plants that had been scoured off more northern terrains. Then, after the glaciers had melted and gone, the climate here began to get warmer and this allowed plants of more southern origin to begin to move north into this same area.

Thus, the pine barrens of New Jersey is a meeting ground for several northern as well as many southern species of plants. McCormick, 1970, stated that 14 basically northern species reach the southern limits of their distributional ranges in New Jersey's pinelands. Most notable of these are **curly-grass fern** and **broom-crowberry,** while several other pinelands plants, such as **pine-barren heather** and **bearberry**, also are of northern origin.

As the climate warmed up here, many plants of southern origin, at least 109, have moved north to reach the northern limits of their distribution in the New Jersey pine barrens. Most of these, such as **pyxie, bog-asphodel,** and **pine-barren gentian**, moved up along the coast, while a few, most notably, **turkey-beard**, moved north along the Appalachian Mountains. Two plant species, **sand-myrtle** and **Pickering's morning-glory**, are believed to occur only in the New Jersey pine barrens, while two other plants, **Knieskern's beak-rush** and **grass-leaved blazing star,** may also occur in similar pine barrens habitats in Delaware.

Of the 14 species of plants of northern origin that reach the southern limit of their distribution in the pine barrens of New Jersey, the following eight species are included in this book.

Conrad's broom-crowberry — *Corema conradii*
Bearberry — *Arctostaphylos uva-ursi*
Yellow pond-lily or spatterdock (to Del.) — *Nuphar variegata*
Pine-barren heather (to Del.) — *Hudsonia ericoides*
Sickle-leaved golden aster — *Chrysopsis falcata*
Seven-angled pipewort — *Eriocaulon aquaticum*
Bog aster (to Del.) — *Aster nemoralis*
Swamp or bog goldenrod — *Solidago uliginosa*

Of the 109 or more species of southern origin that reach the northern limit of their distribution in the pine barrens of New Jersey, the following 42 species are included in this book.

Pyxie — *Pyxidanthera barbulata*
Southern twayblade — *Listera australis*
Swamp-pink (also on Staten Is.) — *Helonias bullata*
Lance-leaved violet — *Viola lanceolata*
Primrose-leaved violet — *Viola primulifolia*
Ipecac spurge — *Euphorbia ipecacuanhae*
Sand-myrtle — *Leiophyllum buxifolium*
Early or flattened pipewort — *Eriocaulon compressum*
Turkey-beard — *Xerophyllum asphodeloides*
Chaffseed — *Schwalbea americana*
Pine-barren sandwort (also on Staten Is. & Long Is.) — *Arenaria=Minuartia caroliniana*
Goat's rue — *Tephrosia virginiana*
Orange milkwort (also on Long Is.) — *Polygala lutea*
Bog-asphodel — *Narthecium americanum*
Swollen bladderwort — *Utricularia inflata*
Pencil-flower — *Stylosanthes biflora*
False or viscid asphodel — *Tofieldia racemosa*
Red milkweed — *Asclepias rubra*
Spreading pogonia — *Cleistes divaricata*
Gold-crest — *Lophiola aurea*
Lance-leaved centaury — *Sabatia difformis*
Crane-fly orchid — *Tipularia discolor*
Pickering's morning-glory — *Stylisma=Breweria pickeringii*

Little ladies' tresses	*Spiranthes tuberosa*
Ten-angled pipewort	*Eriocaulon decangulare*
Short-leaved milkwort	*Polygala brevifolia*
Sclerolepis	*Sclerolepis uniflora*
Slender aster	*Aster gracilis*
Crested yellow orchid	*Habenaria cristata*
Southern yellow orchid	*Habenaria integra*
Grass-leaved blazing star	*Liatris graminifolia*
Narrow-leaved sunflower	*Helianthus angustifolius*
Grass-leaved ladies' tresses	*Spiranthes praecox*
White boneset	*Eupatorium album*
Pine-barren boneset	*Eupatorium resinosum*
Pine-barren gerardia	*Agalinis purpurea* var. *racemulosa*
Wand-like goldenrod	*Solidago stricta*
Late purple aster	*Aster patens*
Pine-barren goldenrod	*Solidago fistulosa*
Silvery aster	*Aster concolor*
Pine-barren or slender rattlesnake-root	*Prenanthes autumnalis*
Pine-barren gentian	*Gentiana autumnalis*

Scientific descriptions of new species are based upon the finding and collection of previously unknown forms, and the locations where these new forms were discovered make up an important component of each original description. Of the more than two dozen species and forms of plants that have been originally described from specimens first found in the pine barrens of New Jersey, the following ten species are included in this book.

Conrad's broom-crowberry	*Corema conradii*
Sand-myrtle	*Leiophyllum buxifolium*
Bog-asphodel	*Narthecium americanum*
Gold-crest	*Lophiola aurea*
Sickle-leaved golden aster	*Chrysopsis falcata*
Pickering's morning-glory	*Stylisma=Breweria pickeringii*
Short-leaved milkwort	*Polygala brevifolia*
Southern yellow orchid	*Habenaria integra*

Grass-leaved ladies' tresses   *Spiranthes praecox*
Pine-barren boneset   *Eupatorium resinosum*

THREATENED AND ENDANGERED SPECIES

Mention needs to be made of threatened and endangered species of plants in the pine barrens. Some species become rare or extinct over time as the result of natural events such as sea-level rise, glaciation, and warm-dry periods. But, the more recent and growing number of threatened and endangered species in the pine barrens is largely the result of human disturbances that usually have an adverse impact upon the natural vegetation.

In the *Comprehensive Management Plan* (CMP) of the New Jersey Pinelands Commission, there is a listing of pine barrens plants that are considered to be either threatened or endangered. For the record, a threatened species is one which may become endangered in the foreseeable future if the environment in the pine barrens deteriorates or other limiting factors are altered. An endangered species is one whose survival in the pine barrens is in imminent danger of extinction.

The following 24 pine barrens species, listed as either threatened or endangered in the CMP, are included in this book.

THREATENED

Southern twayblade   *Listera australis*
Swamp-pink   *Helonias bullata*
Bog-asphodel   *Narthecium americanum*
Red milkweed   *Asclepias rubra*
Sickle-leaved golden aster   *Chrysopsis falcata*
Pickering's morning-glory   *Stylisma=Breweria pickeringii*

Purple bladderwort   *Utricularia purpurea*
Little ladies' tresses   *Spiranthes tuberosa*
Sclerolepis   *Sclerolepis uniflora*
Pine-barren boneset   *Eupatorium resinosum*
Silvery aster   *Aster concolor*

ENDANGERED

Conrad's broom-crowberry	*Corema conradii*
Large or lily-leaved twayblade	*Liparis liliifolia*
Chaffseed	*Schwalbea americana*
False or viscid asphodel	*Tofieldia racemosa*
Spreading pogonia	*Cleistes divaricata*
Yellow fringed orchid	*Habenaria ciliaris*
Reversed or reclined bladderwort	*Utricularia resupinata*
Crested yellow orchid	*Habenaria cristata*
Southern yellow orchid	*Habenaria integra*
Wand-like goldenrod	*Solidago stricta*
Pine-barren or slender rattlesnake-root	*Prenanthes autumnalis*
Pine-barren gentian	*Gentiana autumnalis*

In addition, the following species probably should have been included in the CMP listing, but was not:

Arethusa or dragon's mouth	*Arethusa bulbosa*

Preservation and protection of pine barrens' habitats where these species can still be found are vital to their future survival and every effect must be expended to protect and preserve these critical habitats.

ANATOMY OF A FLOWER
A simplified explanation

Aside from the fact that a flower is a thing of beauty, it really is an important, vital, functioning part of a plant. Its basic purpose is reproduction which it achieves by attracting insects or birds to visit the flower. These bring (male) pollen to a flower from other flowers and then carry some of its own pollen to still other flowers. The resulting cross fertilization of the (female) ovary then produces the seeds or fruit that will germinate and become new plants. In some species of plants, however (ex: pines), this cross fertilization may be brought about by winds.

Most flowers consist of many parts or organs. A few flowers are rather simple and are known as regular flowers (ex: a lily). Many others are more complex, some exceedingly so, such as an iris, members of the pea family, or orchids. Following is a brief review of the more easily observable parts of a flower and the terms that are applied to them.

The flower of a plant is composed of several organs or parts which exist for the express purpose of producing seed or fruit. These flower parts may either function as actual seed producers or they may serve as organs to attract insects and birds to the flower. The essential organs consist of the (female) **pistil**(s) and ovaries, and the (male) **stamens**. When both stamens and pistil(s) are present in the same flower, the flower is considered to be a perfect flower. However, some species of plants produce only male or staminate flowers on one plant and female or pistillate flowers on other plants. Such plants are termed dioecious and a good early-flowering example in the pine barrens is Conrad's broom-crowberry.

Most flowers contain two sets of outer coverings or floral envelopes. The inner of these is the **corolla** which, when fully open, is often (but not always) split into a number of **petals** that are usually (but not always) brightly colored. The outer of these two coverings is the **calyx** which, when the flower opens, is usually split into a number of **sepals** that are usually (but not always) green(ish) in color. If only one set of these coverings is present, they are considered to be sepals even though they may be brightly colored. When a flower is fully opened, and both petals and sepals (together called tepals) are present, the sepals usually are underneath and alternate between the petals.

To return to the essential organs (stamens and pistil), the male **stamen** consists of a stalk or **filament** and, at its tip, an **anther** which bears the pollen. The female **pistil**, or seed-bearing organ, consists of an **ovary** at the base, a **stigma** at its tip, and a style connecting the two. The stigma serves as receptive organ for the male pollen brought from other flowers. The ovary contains the eggs or ovules which, after fertilization, become the seeds or fruit.

Mention should be made of the makeup of composite flowers (ex: asters, dandelions, goldenrods). Composite flowers are made up of two types of flowers in the same flower head. The showy, outer flowers (ex: outer white petals of a daisy) are known as ray flowers. The inner, central disk (ex: yellow center of a daisy) is made up of many closely packed tubular flowers called florets. Some composite flowers, however, consist of only either ray flowers or disk flowers. One other feature of many composite flowers is the presence of modified leaves called bracts which surround the base of the flower head; these bracts may either adhere closely to the flower head or their tips may curl and spread outwards.

GLOSSARY

acute Ending in a sharp point of less than 90°.

alternate Borne singly at each node, first on one side then on the other. Not opposite.

annual A plant that germinates, blooms, and sets seed during a single growing season.

anther The enlarged tip (usually) of a stamen that bears the male pollen.

ascending Rising somewhat obliquely, or curved upward.

axil The upper angle where the leaf joins the stem.

banner The upper, usually enlarged petal of a flower of the pea or bean family; the standard.

bearded Bearing a tuft of long or stiff hairs.

biennial A plant that completes its life cycle in two years.

bloom A whitish, powdery, often waxy coating of a surface.

bract A very small or modified leaf, either growing at the base of a flower (usually) or as a greatly reduced leaf on upper stems or flower stalks.

bud A leaf or flower in the process of development.

bulb An enlarged, underground leaf bud with thick, fleshy scales that serves as a food storage organ.

calyx All of the outer parts of a flower, or sepals, collectively.

capsule A dry fruit that opens when mature to release its seeds.

chlorophyll The characteristic green pigment in the cells of plants that is essential in photosynthesis.

clasping Partially surrounding the stem, as a leaf.

clavate Shaped like a club or a baseball bat, thicker at upper or outer end than at base.

compound (leaf) A leaf divided into two or more separate, smaller leaflets.

cordate Heart-shaped, with a notch at the base and a point at the tip.

corm The enlarged, fleshy, bulb-like base of a stem, but solid, not with fleshy scales. Serves as a food storage organ.

corolla All of the inner, or showy (usually) parts of a flower, or petals, collectively.

corona A set of petal-like structures or appendages between the corolla and the stamens.

creeping Growing along, or just beneath, the surface of the ground, rooting at intervals, usually at the nodes.

deciduous Falling (as leaves) after completion of normal function. Not persistent.

dentate Toothed, usually with the teeth pointed and spreading outward.

dioecious Having male and female flowers, or other reproductive structures, on different plants.

diurnal Pertaining to daytime. Diurnal flowers open during the day.

elliptical In the form of an elipse, broadest in the middle, tapering equally toward both rounded ends.

entire With a continuous, smooth margin (as a leaf), not broken by teeth or lobes.

ephemeral Lasting only a short while. An ephemeral flower lasts only for a day, or less.

evergreen Remaining green throughout a winter.

falcate Sickle- or scythe-shaped. Curved.

fertile Capable of producing seeds or fruit.

filament That part of a stamen that bears an anther.

filiform Long and slender; thread-like.

fleshy Thick and juicy. Succulent.

floret A small, individual flower, usually part of a cluster, as in a composite.

-foliolate Suffix to indicate the number of leaflets in a compound leaf.

fruticose Shrubby.

glabrous Smooth, without hairs.

glaucous Covered or whitened with a fine, waxy powder or bloom.

globose More or less spherical or globular.

head A dense cluster of sessile or nearly sessile flowers crowded closely together at the tip of a flower stalk.

herb A plant with no persistent woody stem above ground, whose stems die back to the ground at the end of each growing season.

hirsute Pubescent, with rather coarse or stiff hairs.

inflorescence The flowering parts of a plant, considered together.

internode The portion of a stem between nodes.

irregular flower A flower whose similar functioning parts, such as petals and sepals, are dissimilar in form.

keel A central ridge, or the two partly united lower petals of many members of the pea or bean family.

lanceolate Lance-shaped, considerably longer than wide. Widest below the middle, tapering to the tip.

leaflet A single division of a compound leaf.

linear Very long and narrow, often with parallel veins and margins.

lip(s) The upper and lower lobes of many irregular, two-lipped flowers, as in the figwort family. Also, the apparently lower, central petal of an orchid.

lobe A projecting segment of an organ, often rounded, too large to be called a tooth.

maculate Spotted, speckled, or mottled.

monoecious Both male and female flowers on the same plant.

naturalized Thoroughly established but originally coming from some foreign area.

nerve A prominent, unbranched, linear vein of a leaf.

net-veined Veins forming a network rather than being parallel veined.

node An enlargement of a stem from which a leaf or a whorl of leaves develops.

oblong Considerably longer than broad, with nearly parallel sides.

obtuse Blunt or rounded at the end.

opposite Situated across from each other at the same node. Arranged in pairs on the stem.

oval Broadly elliptical.

ovary The expanded, basal part of the pistil containing the ovules, where seeds develop.

ovate Egg-shaped, with the broader end at the base.

palate A projection on the lower lip of a flower, constricting or closing the throat.

palmate (leaf) Leaflets radiating from a central point.

parallel-veined With several or many more or less parallel veins.

parasite A plant that obtains its food from another living plant, to which it is attached.

parted Deeply cut, usually more than half way to the midvein or base.

pedicel The stalk of a single flower in an inflorescence.

peduncle A primary flower stalk, supporting either a cluster or a single flower.

perennial A plant that usually lives more than two years, often living for several years.

perianth The floral envelope, consisting of all of the sepals and petals (or tepals) of a flower, collectively.

persistent Remaining attached, as leaves through a winter.

petal One of the inner set of floral parts, often colored or white, that serves to attract pollinators. One of the segments of a corolla.

petiole The support or stalk of a leaf.

pinnate (leaf) Leaflets arranged on each side of a common axis.

pistil The female, or seed-bearing organ of a flower, usually consisting of an ovary, a style, and a stigma.

pistillate flower A flower with one or more pistils, but no stamens.

pollen The male spores, or pollen-grains, produced by the anther.

procumbent Growing flat on the ground or trailing, but not rooting at the nodes.

prostrate Lying flat upon the ground.

pubescent Covered with hairs, especially if short, soft, and downy-like.

punctate Dotted, usually with small pits or depressions.

raceme A more or less elongated inflorescence.

ray(s) The flat, outer, petal-like blades that encircle central disk flowers, as in a daisy.

reflexed Bent or curled backward, sometimes abruptly.

rhizome A creeping prostrate or underground stem, often rooting at the nodes. A rootstalk.

rib A prominent, usually longitudinal, vein of a leaf.

rosette A cluster of leaves or other organs arranged in a circle or disk, often in a basal position.

rugose Wrinkled.

saprophyte A plant or a fungus that gets its food from dead, organic matter.

scale A tiny, colorless leaf found on some plant stems.

scape A naked, leafless flower stalk that arises from ground level.

seed The ripened ovule of a plant, capable of germinating into a new plant.

sepal One of the outer set of floral parts, typically green or greenish, and more or less leafy in texture. One of the segments of the calyx.

serrate Toothed along the margin with sharp, forward-pointing teeth.

sessile Attached directly to a base, without a stalk of any kind.

seta A bristle.

shrub A low, perennial, woody plant, usually with several stems.

spathe A large, usually solitary, bract that usually encloses a flower cluster or inflorescence.

spike An elongated flower cluster arranged along a central stem, with individual flowers sessile or nearly so.

spur A hollow, tubular extension of some part of a flower.

stalk The support stem of a single leaf or flower.

stamen The male organ of a flower, consisting of a slender stalk (filament) and a pollen-bearing tip (anther).

staminate flower A flower with one or more stamens, but no pistil.

standard The uppermost petal of a member of the pea or bean family. The banner.

stem The main stem or trunk of a plant.

stigma The part of the pistil that is receptive to pollen for fertilization.

striate Marked with fine, more or less parallel lines.

style The slender stalk of the pistil that connects the stigma to the ovary.

sub- Latin prefix that means under, or almost, or not quite.

succulent Fleshy and juicy, or a plant that has lots of water in its fleshy stems and leaves.

tepal Term used for sepals and petals of similar form.

tomentose Densely pubescent with matted, woolly hairs.

trailing Prostrate on the ground, but not rooting.

tuber A thickened, short, underground stem having buds or eyes.

vein One of a network of tiny channels in a leaf for the conduction of plant fluids.

viscid Sticky or greasy.

weed An aggressive plant that intrudes where it is not wanted.

wing One of the two lateral petals in a flower of the pea or bean family.

WILDFLOWERS

March. 16

April 20

May 30

June. 60

July 102

August 138

September 152

CONRAD'S BROOM-CROWBERRY

Corema conradii Crowberry family

Mid- to late March through mid-April
Not over 2"- 6" high (in NJ)
Branches up to 12" long

A very low-growing, woody sub-shrub with multiple, gnarled branches that spread out in roundish patches over the bare sands and are covered with tiny, needle-like, evergreen leaves. In early spring, this almost prostrate sub-shrub appears very drab, yet tiny flowers appear on the tip ends of last season's growth. After blossoming, the plant produces a new growth of fresh, bright green leaves. Male and female flowers appear as terminal heads on separate plants and are so small that to examine them closely one must use a hand lens. The most easily seen parts of these flowers are the purplish anthers of the male flowers, which stand out rather clearly when in full bloom.

Restricted in New Jersey to pine plains areas where, in places, it grows in abundance. Its northward distribution is scattered, with stations in the Shawangunk Mountains, Ulster County, NY, along coastal Massachusetts and Maine, and then Nova Scotia, Prince Edward Island, and Newfoundland in Canada. Broom-crowberry reaches its greatest size, up to 2' in height, and greatest abundance in Labrador.

PYXIE

Pyxidanthera barbulata Diapensia family

Very late March into early May
Not over 1"- 1 1/2" high
Branches 6"- 10" long

This very small, low, perennial sub-shrub grows flat on the ground. Its numerous, somewhat woody, branching, trailing stems creep along the sand and bear tiny, needlelike, evergreen leaves so crowded together at the ends of the branches that they form thick, moss-like mats which give rise to this plant sometimes being called pyxie moss. However, this is not a moss but a true flowering plant, having a central root system with stems, leaves, flowers, and seeds. In early spring, leaves may be quite reddish, but these turn green by the time the plant blossoms. Numerous small, solitary flowers, about 1/4" wide, usually white or rarely pale pink, are borne at the ends of the short branches, with each flower having five blunt, spreading lobes. Little can compare with the beauty of these white, star-like blossoms that sparkle up from their green, moss-like setting.

Frequent to common throughout open, dry or slightly damp, white sandy woodlands. Abundant in parts of the pine plains.

CONRAD'S BROOM-CROWBERRY

CONRAD'S BROOM-CROWBERRY

D.H. ("Bart") Ulmer, MD

PYXIE

GOLDEN CLUB
Orontium aquaticum Arum family

Very late March through May
Flower spikes 1'- 1 1/2' above water surface

A perennial, aquatic herb, rooted in mud under the water, that produces elliptical leaves that may rise as much as a foot above the water surface but often just float on the surface. The upper surfaces of these leaves are covered with a satiny coating that repels water so that beads of water form on their surfaces before running off the edges of the leaves. This characteristic is the reason for this plant's other common name of "never-wet." Undersides of these leaves are whitish with many parallel veins. Flowers develop on the upper ends of long, underwater stems and appear on naked spikes that rise on white stalks just above the water surface, the spikes themselves covered with tiny, bright, yellow to orange-yellow flowers.

Common to abundant in shallow ponds, reservoirs, bogs, and sluggish streams.

TRAILING ARBUTUS or MAYFLOWER
Epigaea repens Heath family

Very late March to early May
Not over 1 1/2"- 2 1/2" high
Branches 4"- 12" long

This small, low, perennial sub-shrub grows flat on the ground, with slightly woody, somewhat rough-hairy, light brown, trailing stems that creep over the terrain, sometimes forming patches of considerable size. Its oval, thick, 1"- 3" leaves are hairy, leathery, and evergreen, but become smooth above when mature. New stems and leaves are especially hairy. Clustered inconspicuously at the tips of its branches, between the rough, textured leaves of the previous year, small, fragrant, waxy appearing flowers develop that are tubular, quite hairy within, and that expand into five small, flared lobes. Flowers are usually white to pinkish-white or, less frequently, a beautiful pinkish color.

Occasional to frequent in dry, sandy woodlands. Common in parts of the pine plains where the heat of the midsummer sun tends to curl the leaves of the plant until its edges become somewhat brown and brittle. The mayflower name was given to this plant by the Pilgrims, after the name of their vessel, because this was the first flower to greet them after their first, long, fearful winter.

March 19

GOLDEN CLUB

TRAILING ARBUTUS or MAYFLOWER

Doris Boyd

LEATHERLEAF or CASSANDRA
Chamaedaphne calyculata Heath family

Early April to mid-May
1'- 3' high

A low, erect shrub with profuse, tough, wiry branches. Leaves small, alternate, leathery but thin, partially evergreen, oblong-elliptical in shape, and covered on both sides, more so beneath, with small, roundish, rusty scales, especially when young. Upper leaves very small, almost reduced to bracts. Flowers white, fragrant, and bell- or urn-shaped, forming long, one-sided rows of hanging flowers out to the tips of the branches. Witmer Stone, an early southern New Jersey botanist and ornithologist, and author of the classic *The Plants of Southern New Jersey*, 1910-1911, wrote that "its one-sided racemes of white, cylindrical flowers ... have gained for it the name of 'false teeth brush.'"

Abundant in swamps, bogs, peaty swales, and along the edges of old cranberry bogs. This is an invasive species that, if allowed, will outcompete most other vegetation in its habitat.

SWAMP-PINK
Helonias bullata Lily family

Early April to mid-May
Flower heads 1'- 2 1/2' high

This perennial lily is one of the earliest terrestrial herbs to bloom in the spring. From a basal rosette of last year's dark, olive-green leaves, now lying flat on the ground, new, light green leaves develop for the coming season, eventually reaching a height of about 1 foot. At almost the same time, from the center of some of these leaf clusters, one or two hollow, nearly leafless scapes arise to each bear a beautiful, oval to oblong cluster of 40-60 tiny, bright, fragrant flowers. Each flower possesses six lavender-pink tepals (petals and sepals), protruding from which are six stamens tipped by vivid blue, pollen-bearing anthers. Swamp pink is one of the most beautiful of all flowering plants in southern New Jersey.

Local in both hardwood and cedar swamps and bogs, usually found along or near the edges of small, running streams.

April 21

LEATHERLEAF or CASSANDRA

SWAMP-PINK

SOUTHERN TWAYBLADE
Listera australis Orchid family

Early or mid-April to mid-May
4"- 10" high

A perennial herb with two ovate, rather egg-shaped, smooth, shining leaves located slightly above the middle of its stem. At the top of this stem is a loose spike of 6-15 small, yellowish-greenish-purplish flowers. Both petals and sepals are minute but the lip of the flower is 1/4"- 1/2" long, almost spade-like, deeply split, and at least four times as long as the tiny petals.

Rare in low, damp, well-shaded woodlands. Because this orchid is so small, and because its inconspicuous flowers blend so well with dead leaves on the forest floor, it is a very difficult flower to spot in the woods. Extremely rare in southern New Jersey. This writer knows of only two locations for it: one well within the pine barrens and one near the coast.

BIRD'S-FOOT VIOLET
Viola pedata Violet family

Mid- or late April to mid-May
4"- 10" high

Most violets are low-growing herbs with leaves and flowers arising from a rootstalk. Whereas many violets have entire, often heart-shaped leaves, bird's-foot violets have smooth leaves that are divided into three to five segments, hence its name from the assumed resemblance of its leaves to a bird's foot. Often, some segments are again cut and toothed so that the average leaf may be deeply divided into nine or more narrow points. Flowers are fairly large and showy, from 3/4"- 1 1/4" across, with five petals: two upper ones, two lateral ones, and one lower petal that is wider than the others. The petals of most pine barrens bird's-foot violets are all the same light blue-violet to light purple-violet in color, although the lower petal may be whitish, veined with violet. Conspicuous in the center of each flower is a cluster of orange-tipped stamens.

This may not be a true pine barren species, having apparently been introduced into the pine barrens. Most commonly seen along dry roadsides.

SOUTHERN TWAYBLADE

BIRD'S-FOOT VIOLET

LANCE-LEAVED VIOLET
Viola lanceolata Violet family

Mid- or late April to mid-June
2"- 6" high

    The smooth, narrow leaves of this violet are lance-shaped, 2"- 6" high, but only 1/2"- 3/4" in width, tapering down to a very narrow stalk. This white violet has five white petals but the lower three petals are striped with purplish veins. This violet is considerably more common than the primrose-leaved violet, below, and is considered to be the most abundant violet of any kind in the pine barrens.

    Common in moist, open, sandy soils.

PRIMROSE-LEAVED VIOLET
Viola primulifolia Violet family

Mid- or late April to mid-June
2"- 8" high

    This white violet blossoms at the same time as the lance-leaved violet, above, but can easily be separated from the above by its wider, oval or egg-shaped leaves and by the abrupt narrowing of the leaf to its stalk. Flowers of the two species are nearly identical (not illustrated).

    Occasional in moist, open, sandy soils.

IPECAC SPURGE
Euphorbia ipecacuanhae Spurge family

Mid- or late April to late May
6"- 12" high

    This low-growing, perennial herb is best known for its differently colored leaves. From a deep, thick, central rootstalk, several repeatedly branching, smooth stems produce an abundance of somewhat fleshy leaves that vary greatly in arrangement, size, shape, and color. Most lower leaves are opposite, while upper ones may be in small rosettes. Some leaves are quite linear, others broadly oval, with all possible intermediates. Leaves on some plants are quite green, while on others they are deep maroon or reddish-purple, even when growing side by side. Occasionally, both red and green leaves will be found on the same plant. Both leaves and stems are filled with a milky juice characteristic of the spurge family. The plant's minute, yellow-green flowers, which do not have any petals, are set out above tiny greenish bracts, and bloom *before* any leaves appear.

    Common on open, bare, white sands and along dry, sandy roadsides. This is not the same plant that is the source of emetine, used commercially as an emetic. That plant is a tropical South American creeping plant, *Cephaelis ipecacuanha*, of the Madder family.

LANCE-LEAVED VIOLET

IPECAC SPURGE

BEARBERRY
Arctostaphylos uva-ursi Heath family

Mid- or late April to mid-May
Not over 2"- 8" high
Branches up to 18" long

A low-growing, trailing, and woody sub-shrub with thin, reddish bark. Its many branching stems trail over the ground and bear numerous small, evergreen leaves that are shiny, thick, entire, and round-tipped or paddle-shaped. In places, the foliage develops into such thick mats that the plant forms a veritable carpet over the hot sands. Flowers of bearberry develop in small terminal clusters of hanging flowers that are bell- or urn-shaped with small lobed openings, as is typical of many heath family members. The flowers are waxy-white or pinkish-white or, frequently, white, tipped with deep pink. By mid-August, fertilized flowers have developed into dry, mealy, bright red berries that persist over winter.

Frequent to common throughout the pine barrens, especially along sandy trails and roadsides, and in the pine plains. The bright red berries of bearberry are presumably relished by bears, hence the name: *arcto* and *ursus*, Greek and Latin for "bear"; *staphylos*, Greek for a cluster (of berries); and *uva*, Latin for "grape", thus "bear-cluster, the grape of a bear."

HIGHBUSH BLUEBERRY
Vaccinium corymbosum Heath family

Mid- or late April to late May
3'- 12' high

A tall, bushy, branching shrub. Bark on main stems and branches grayish-brown, with bark on older stems often shredding off to give stems a mottled look. Leaves elliptical, pointed, usually entire, without teeth, green and smooth above, slightly paler green and somewhat hairy beneath, without resin dots. Flowers bell- or urn-shaped, pendant, usually white or slightly pinkish. Fruits blue to blue-black, with a bloom.

Frequent to common in wet thickets and swamps. This is the base stock for the cultivated highbush blueberry of commerce.

April 27

BEARBERRY

HIGHBUSH BLUEBERRY

SAND-MYRTLE
Leiophyllum buxifolium Heath family

Mid- or late April through June
6"- 36" high

This low, spreading, woody shrub, with scraggly stems and branches, rough brown bark, and small, leathery, evergreen leaves appears, at first glance, to be a dwarf variety of English boxwood. The oval or oblong leaves of this odd little bush are a lustrous dark green, with slightly curled back margins, and are crowded near the ends of its branches. Its numerous white or pinkish flowers develop in clusters at the ends of its branches, and the individual flowers each have five petals with 10 purple anthers. These flowers seem to be very attractive to several species of insects, particularly beetles and small butterflies, and occasionally some rare species are found on these flowers.

Frequent, even abundant in places, such as low, damp, sandy, open, pitch pine woodlands. This is one of the most characteristic of all pine barren plants and is typical of the heart of the pines.

BLUE or OLD-FIELD TOADFLAX
Linaria canadensis Figwort family

Mid- or late April to early July
4"- 24" high

A very slender, smooth annual or biennial herb with a few small, linear, light green leaves that are toothless, stemless, smooth, and shiny. Pale blue to pale violet flowers develop in a narrow spike, with the throat of each flower closed by a ridge or "palate" rising from the lower lip, while the rear of the flower is prolonged backward and downward by a hollow extension or spur. Individual flowers are small, about 1/3" long, and two-lipped. The lower lip is large and three-lobed, with a conspicuous white, convex, two-ridged "palate." The upper lip has two acute divisions. The spur is thread-like and slightly curved.

Common. Often found in large colonies in dry, sandy soils of old fields and other disturbed areas. A weed rather than a "good" pine barren species.

April 29

SAND-MYRTLE

BLUE or OLD-FIELD TOADFLAX

WILD LUPINE
Lupinus perennis Pea or Bean family

Very late April to mid-June
8"- 24" high

Lupine is one of the more conspicuous wild flowers in the pine barrens and may occasionally develop into large patches. From a perennial root, palmately compound leaves develop in a wheel-shaped pattern at the tops of their stems, with 7-11, usually eight, 1"- 2" slightly hairy leaflets. Bright purplish-blue flowers develop on 4"- 10" terminal spikes, each flower 1/2"- 2/3" long, in typical pea family form with an upper petal (standard), two lateral petals (wings), and a lower petal (keel).

Occasional in open, dry, sandy soils, and along edges of woods and roadsides.

PINK LADY'S-SLIPPER or MOCCASIN-FLOWER
Cypripedium acaule Orchid family

Very early May into early June
6"- 18" high

A perennial herb with a beautiful flower that rises on a naked scape from a ground-level pair of large, thick, dark green, basal leaves, with conspicuous, unbranched veins the length of each leaf. An unusual feature is that these leaves grow right out of the ground, appearing to be without stalks. The single flower is borne on top of a bare, leafless scape, topped by a single green bract that arches forward over the flower. The principal part of the flower is a large, drooping, hollow pouch (lip or "slipper"), with a narrow slit or cleft down its entire front length. This deeply cleft, pink with reddish veins, or rarely white, pouch is distinctive. If an insect, usually a bee, lands on the cleft, it drops into the sac and the only way out is through two openings at the rear (top) of the pouch. In passing through these openings, the insect brushes against the reproductive organs, thus insuring pollination. Conspicuous above and behind the pouch are six widely spreading, greenish-brown sepals and petals, the two lower sepals usually grown together.

Frequent in dry or moist, acid woods, often among pines and oaks. This is the only lady's-slipper to be found throughout most of the pine barrens.

April–May 31

WILD LUPINE

PINK LADY'S-SLIPPER or MOCCASIN-FLOWER

STARFLOWER
Trientalis borealis Primrose family

Early May to early June
2"- 10" high

A small, delicate, woodland plant with a long, horizontally creeping root that sends up thin, almost bare stems, each terminating in a whorl of five to nine shiny, lance-shaped, light green leaves that taper to both ends. From the center of this whorl rise one or two thread-like stems, each bearing a single, fragile, star-shaped, white flower with five to nine, usually seven, pointed petals. The stamens are long and delicate, with tiny golden anthers.

Occasional and local along the edges of open cedar bogs and thin, moist woodlands.

BLACK HUCKLEBERRY
Gaylussacia baccata Heath family

Early May to mid-June
1'- 3' high

A low, erect shrub with stiff, brown branches. Its oval-oblong, pointed leaves are 1"- 2" long, yellow-green, deciduous, and covered on both surfaces with shiny, orange resin dots, especially on young leaves. These dots can best be seen on the undersides of young leaves using a hand lens. Flowers develop in small, pendant, one-sided clusters, are urn-shaped like other heath family members, and are brick-red in color. The berry-like fruits that develop later are black.

Common to abundant in dry or moist thickets and woods throughout the pine barrens. This may be the most common, most abundant member of the heath family in the pines. In fact, it is almost ubiquitous.

May 33

STARFLOWER

BLACK HUCKLEBERRY

STAGGERBUSH
Lyonia mariana Heath family

Early May through June
1' to 4'- 6' high

A shrub with erect, or nearly erect, smooth, wand-like branches. Flowers develop on the leafless branches of the previous season before new leaves appear. When leaves do develop, they are large, 1 1/2"- 2 1/2" long, oblong-elliptical in shape, olive-green, smooth above, slightly hairy on the veins underneath, and deciduous. Flowers are large, up to 1/2" long, and white, sometimes tinged with pink, and appear as small, pendant clusters of urn-shaped, nodding blossoms. Witmer Stone, in *The Plants of Southern New Jersey*, 1910-1911, wrote that these "flowers ... are the largest and handsomest of any of the urn-shaped blossoms" of the huckleberries and other heath family members. Unlike huckleberries, however, flowers of staggerbush do not develop into edible fruits but, instead, develop into seed-containing capsules.

Common to sometimes abundant in low, moist, sandy soils, thickets, and peaty swamps, and along damp, sandy trails. This shrub gets its common name from the belief that it is poisonous to cattle and makes them stagger.

DWARF DANDELION
Krigia virginica Aster or Composite family

Early May to late June
2"- 12" high

A small, annual herb with one to several slender, leafless flower stalks that rise from a basal rosette of irregularly lobed and toothed, dandelion-like leaves. Later in the year, the plant may become branched and develop stem leaves. Each flower stalk bears a single, small, ray-type, hawkweed-like golden- or orange-yellow flower, scarcely more than 1/4"- 1/2" wide.

Frequent in dry, sandy soils, especially in disturbed areas. One of the earliest of the relatively few composite-type flowers to bloom in the spring.

May 35

STAGGERBUSH

DWARF DANDELION

FROSTWEED

Helianthemum canadense Rock-rose family

Early May to late July
6"- 24" high

A small, perennial herb with erect stem and alternate, simple, entire, narrow, dull green leaves. Both its stem and its leaves are covered with fine hairs, the leaves especially so on their undersides. This plant produces two types of flowers, dependent upon the season. Spring flowers are solitary, terminal, and large, 1" or more across, with five showy, yellow, wedge-shaped petals and innumerable stamens. Petals of these flowers fade early and soon fall away, so that each of these blossoms lasts for only a single day. Later in the season, July into early September, after the spring flowers have passed, the branches of this plant elongate and smaller flowers, without petals, develop at the bases of the leaves, but these flowers do not open. Both types of flowers produce seed capsules, those of the spring flowers about 1/4" in length, those of the later flowers barely larger than a pinhead.

Occasional in dry, sandy soils and along similar roadsides. This plant is remarkable for the fact that, in late autumn, ice crystals sometimes form in the cracked bark of its lower stem and roots.

EASTERN BLUE-EYED GRASS

Sisyrinchium atlanticum Iris family

Mid-May to mid-June
6"- 24" high

A small, slender, perennial with stiff, narrow grass- or iris-like leaves and flowering stems that are flat, sharp-edged, and taller than the leaves. At least some of the flowering stems may be branched. Often grows in tufts or clumps that may be mistaken for a grass when the plant is not in flower. Flowers are small and delicate, have six spreading petals (technically three petals and three sepals) all alike, each tipped with a tiny bristle, and all are blue-violet in color with yellow centers. Thus, the blue-eyed name for this species is misleading, for it is actually yellow-eyed. Flowers open only on bright, sunny days, and each one remains open for just one day.

Common in moist, grassy fields, wet meadows, and marshes.

FROSTWEED

EASTERN BLUE-EYED GRASS

PIPEWORTS
Eriocaulon spp. Pipewort family

Pipeworts are wetland herbs that grow either with their leaves under water and their flowering heads standing above water level, or with the entire plant out of water. These plants develop from tufts of fibrous roots which have cross-lines in them so that they may look like tiny, whitish angleworms. Leaves are all basal and develop in clusters. Leaves are thin, narrow, smooth, parallel-veined, and grass-like, and taper from being wider near their base to long, thread-like tips. Leaves are usually checkerboarded with fine veins so that they may appear transparent or window-like. Flowers are borne on naked stalks that are sheathed at the base. Flowers consist of dense, round heads of very small, whitish or grayish-white flowers, the whole resembling old fashioned hatpins.

There are three species of pipeworts in the pine barrens. All grow in wet habitats such as bogs, shallow ponds, marshes, wet grasslands, and savannas.

EARLY or FLATTENED PIPEWORT
Eriocaulon compressum Pipewort family

Mid-May to late June
6"- 20" high

This small, tufted herb develops as a cluster of short, sheathed, linear, spreading, basal leaves, 2"- 4" long, that gradually taper to a sharp point. From this clump rise one or more slender stalks, each topped by a solitary, compact head of woolly, grayish-white flowers approximately 1/2" across. Its early flowering period and short leaves easily separate this species from the two later blooming pine barren pipeworts with the same type of button-shaped, grayish-white flowers.

TEN-ANGLED PIPEWORT
Eriocaulon decangulare Pipewort family

Mid-July to early October
1'- 3' high

This is the largest, most conspicuous, and may be the most commonly seen of the three pipeworts found in the pine barrens. The most obvious difference between this and the earlier seven-angled pipewort (see page 40) is size. Also, this begins to bloom about two weeks later than the seven-angled pipewort. Here, the flower stalks reach 1'- 3' high, are 8-12-angled or ribbed (use hand lens), and bear the same type of woolly, grayish-white flowers. (not illustrated)

EARLY or FLATTENED PIPEWORT

SEVEN-ANGLED PIPEWORT

Eriocaulon aquaticum = septangulare
  Pipewort family

Early July to early October
1"- 8" high

This very small, tufted pipewort has 1"- 3" long, transparent leaves with window-like openings in their veins. In submerged specimens, leaves may become better developed, up to 1' in length, and may float. From this clump of leaves, a weak, twisted, flower stalk rises 1" or 2" to possibly 8", and is usually four- to seven-angled or ribbed (use hand lens). On top of this flowering stalk is a small, about 1/8", roundish head of woolly, white flowers, interspersed with numerous leaf-like bracts. This species does not begin to bloom until after the earlier, flattened pipewort has finished blooming and is much smaller that the ten-angled pipewort that will begin to bloom by mid-July. This may be the least conspicuous and most difficult to find of the three pipeworts in the pine barrens (not illustrated).

TURKEY-BEARD

Xerophyllum asphodeloides Lily family

Mid-May to early July
1'- 5' high

A perennial herb with a thick, basal clump of dry, wiry, grass-like leaves, 1' or more in length, that remain persistent throughout the year. From this rises a tall flower stalk with some shorter, stiff leaves ascending the stalk. Topping off this stalk is a dense cluster of many small, white, six-pointed flowers, the cluster itself being as much as 3"- 10" tall.

A characteristic pine barrens plant that is frequent to common in low, sandy pinelands. One of only two species in its genus, the other being beargrass, *Xerophyllum tenax*, found in Glacier National Park, Montana, and vicinity.

YELLOW STAR-GRASS

Hypoxis hirsuta Lily family

Mid-May to the end of August
Leaves 2"- 12" high
Flowers 2"- 6" high

A low, perennial herb with a tuft of narrow, grass-like, basal leaves that rise from a thick, underground stem. Flowering stems are erect and slender, usually shorter than the leaves, and both leaves and flowering stems are covered with hairs. From one to seven, more often two or three, yellow flowers bloom in a small cluster at the top of each flower stalk. Each individual flower is composed of three petals and three sepals, but since these are the same color and size, each flower is six-pointed and star-like. Although each flower is deep yellow inside, it is greenish and hairy on the outside.

Frequent in either dry or moist, acid, sandy, grassy soils, including roadsides.

May 41

TURKEY-BEARD

YELLOW STAR-GRASS

YELLOW POND-LILY or SPATTERDOCK
Nuphar variegata Water-lily family

Mid-May through September
Many leaves float on water surface
Flowers 2"- 3" above water surface

This pond-lily has large, up to 10", broadly oval, thick, dark green leaves that rise on flattened stems from tuberous roots down in the mud. Each leaf has a deep, narrow, V-shaped notch, with rounded lobes, and while some leaves may stand up a bit above the surface of the water, most leaves appear to be simply floating on the water's surface. The conspicuous part of a yellow pond-lily is the round, globe-like, yellow flower, open at the top, that is composed of five or six greenish-yellow sepals. Centered conspicuously inside this globular flower is a disk-like stigma, that part of the pistil that is receptive to pollen.

Frequent to common in water at the edges of ponds and streams and in swamps.

There is another, similar species of yellow pond-lily, *Nuphar advena*, in southern New Jersey. Botanical differences between the two are small. Leaves of *N. advena* usually rise and stand erect above the water surface and the inner sepals of the flower are yellow, tinged with purple. The leaves of *N. variegata* have flat stems, mostly lie on the surface of the water, and the inner sepals of the flowers tend to be reddish. *N. advena* is more frequently encountered along the lower Delaware River and its adjacent tidewater streams, while *N. variegata* is more characteristic of the pine barrens.

YELLOW POND-LILY or SPATTERDOCK

PUTTY-ROOT
Aplectrum hyemale Orchid family

Mid- or late May to very early June
10"- 16" high

    Solid, tuberous roots of this perennial herb first send up one or more leafless, flowering scapes containing 7-15 yellowish- or purplish-brown flowers, each about 1/2" long. Lips of these flowers are whitish with purplish markings and are crinkly-edged. The flowering period of this orchid is very brief. There are no leaves at this time.

    Later, in early fall, this plant produces a single, broad, oval, wrinkled, 4"- 6" leaf with distinct folds and whitish lines. This leaf persists over the winter but perishes by the time the new flowering scapes develop in the spring.

    Rare in rich, wooded soils, and, because its inconspicuous flowers blend in so well with dead leaves on the forest floor, this is a very difficult flower to spot in the woods. This is not generally regarded as a pine barren species. However, it has been reported from the Wharton State Forest, all of which is in the pine barrens, and this photograph was taken in the Mantua Creek watershed in Gloucester County, just beyond the western fringes of the pine barrens.

PUTTY-ROOT

PINE-BARREN or GOLDEN HEATHER
Hudsonia ericoides Rock-rose family

Mid-May to early June
4"- 8" high

This small, bushy, multiple-branching, perennial sub-shrub has somewhat woody branches densely covered with tiny, bristly, needle-like, green leaves that spread slightly outward from the stem. Small, bright yellow flowers appear on hairy stalks from the tips of the branches, sometimes so densely as to almost conceal the foliage. When in bloom, great patches of this low shrub may look like a massive carpet of bright, golden bloom. Individual flowers open in sunshine and last three or four days, but the overall blossoming period is quite brief.

Common to abundant on dry, white, sandy stretches. A characteristic pine barrens species.

BEACH or WOOLLY HEATHER
Hudsonia tomentosa Rock-rose family

Mid-May to early June
4"- 8" high

This very similar species is sometimes found growing side by side with golden heather in the pine barrens. The main difference between the two is that the tiny, awl-shaped leaves of the beach heather are scale-like, soft-woolly, grayish-green, somewhat hoary, almost overlap each other, and remain close to their stem. Flowers of beach heather develop on very short stalks or none at all (not illustrated).

Occasional and local in the same pine barren habitats as golden heather, flowering at nearly the same time or slightly later. Primarily a coastal dune species where it may become abundant.

Although called heathers, neither of these are, in fact, a heather. The true heather is a European species.

PINE-BARREN or GOLDEN HEATHER

ARETHUSA or DRAGON'S MOUTH

Arethusa bulbosa Orchid family

Mid- or late May to mid-June
5"- 12" high

The smooth scape that rises from a small bulb bears two or three small bracts, with the topmost one later developing into a single, narrow leaf after the flower has finished blooming. Rarely, a bulb may produce two such scapes. The single, delicate flower that develops at the top of this scape is a combination of magenta-pink sepals and lateral petals, up to 2" long, that stand almost erect, their ends arching, hood-like, over the column. The broad, lower lip in front bends down somewhat, is suffused with pink splotched with darker pink or crimson-purple, bears rows of yellow (usually) or whitish hairs, and has a ragged or fringed margin. This lip is known as a "beard" and serves as a landing platform for insects, usually bumblebees.

Local and often rather rare in bogs, peaty meadows, and damp, sphagnum depressions. Arethusa is the first of three closely related bog orchids that come into bloom in quick succession toward the end of May and into June. The other two are rose pogonia and grass-pink, descriptions of which appear on pages 60 and 62. All three are exquisitely beautiful wild flowers.

SLENDER BLUE FLAG

Iris prismatica Iris family

Mid-May to late June
Leaves up to 2' high
Flowers up to 3' high

A perennial herb with a tuberous, thickened rootstalk from which rise two or three very narrow, almost grass-like leaves. The slender, irregular flowering stalk rises above the leaves and bears one or two violet-blue flowers. These flowers have dark, purplish veins on the light yellowish bases of the three downward curved sepals, which are larger and more conspicuous than the three erect petals in the center of the flower.

Frequent in pine barren swamps and open, sunny, wet, boggy marshes, meadows, and ditches.

ARETHUSA or DRAGON'S MOUTH

SLENDER BLUE FLAG

Doris Boyd

PITCHER-PLANT
Sarracenia purpurea Pitcher-plant family

Late May to mid-June
8"- 24" high

Pitcher-plants are perennial, carnivorous herbs with basal rosettes of smooth, tubular leaves, 4"- 6" high, that are modified as traps and arise from stout, fibrous roots, often embedded in sphagnum. These leaves are shaped like tubular pitchers that are open to the sky, hold water, and are persistent over winter. At the tip of each leaf is a lip that is covered with inward- and downward-pointing, bristly hairs. Each leaf is keeled toward the inner side of the rosette and toward the central flower stem. Pitchers in open areas may be colored from green to shades of yellow or red and usually are strongly veined with red-purple, but leaves of plants that grow in deep shade, as in cedar swamps, are often just plain green. Insects and other prey are attracted to these pitchers by their colors, their scents, and their secretions. After landing on a lip, insects often fall into the pitcher, and, since they can neither fly nor crawl out, they die by drowning and soon decay. The ability of these and other carnivorous plants, such as sundews, to attract animal prey and to cause their deaths is an adaptation that helps to provide these plants with nutrition, through absorption of minerals from the decayed bodies of insects, into the tissues of the plants. This supplements their other food sources and helps them to better survive in the acid, nutrient-poor soils that they inhabit.

Flowers of pitcher-plants develop singly at the tops of tall, slender scapes. As a flower bud approaches its opening, its stem bends like a shepard's crook and the nodding flower opens facing down. These flowers are round, 1 1/2"- 2" across, and range from green to yellowish to, often, deep reddish-purple. The sepals, on top, are madder-purple on the top- or outside, but greenish on the inner- or underside. Five maroon petals hang down from the sides of the flower like little drapes, and inside these drapes the smooth, round face, or central disk, of the flower is green. It is behind this central disk, called an umbrella, where the reproductive organs of the flower are located. Later, after the petals drop, the flower heads usually assume a nearly erect position. In time, what remains of the flowers dry up and the remaining brown, round, tubular seed pods stand like mute sentinels above the open bog.

Common to abundant in peaty bogs, cedar bogs, swamps, and open cedar savannas. Though often referred to as insectivorous plants, the correct term for all such plants is carnivorous because some plants are known to entrap larger forms of animal life, such as small water animals that clearly are not insects, and spiders, slugs, snails, small amphibians such as frogs, and even small birds.

May 49

PITCHER-PLANT (blossom)

PITCHER-PLANT (cluster of leaves)

PARTRIDGE-BERRY
Mitchella repens Madder family

Mid- or late May to late June
Not over 1"- 1 1/2" high
Branches 6"- 15" long

This small, creeping, vine-like herb grows flat on the ground and has slender, trailing stems that are freely rooting at the nodes. Leaves are paired and opposite, oval or roundish, blunt at apex, possibly somewhat heart-shaped at base, and evergreen. Leaves are green with white veining along the midrib. Flowers, which are creamy-white or slightly pinkish, waxy-looking, and fragrant, are borne in pairs at the ends of the branches or from outer leaf axils. Flowers are tubular, about 1/2" long, with four spreading lobes, and are densely fine-hairy on the inside. Sometimes, the two flowers of a pair are partly united, as may be the fruit, which is a scarlet berry that is persistent over winter and is edible.

Occasional in low, damp woodlands.

MOUNTAIN-LAUREL or CALICO-BUSH
Kalmia latifolia Heath family

Mid- or late May to late June
3'- 15' high (in NJ)

A large, woody shrub with rigid, spreading branches, sometimes growing in thick clusters, the trunks and larger branches of which are irregular and angular, with dark, ruddy, brown bark. Leaves, crowded at the ends of the branches, are thick, leathery, entire with smooth margins, pointed at both ends, 2"- 5" long, shiny dark green above, slightly paler underneath, and evergreen. Flowers are borne in large, terminal, dome-shaped clusters, with each individual bloom positioned on the end of a sticky, hairy stalk. Individual blossoms are cup-shaped, with five outer lobes and 10 arching stamens, each with their anthers held securely in tiny pockets in the sides of the flower, or corolla. When visited by a bee, these stamens spring out from their pockets, and their pollen gets sprayed over the insect, which it then carries to the next flower. Unopened, cone-shaped flower buds are often pure pink. Flowers waxy-white, tinged with pink, when fully open.

Common in both dry and moist pine and oak woodlands. May grow to heights of 30'- 35' in the South. The name calico-bush has special significance for the pine barrens. From approximately 1808-1834, there was a small village near Martha Furnace, in Burlington County, that was named Calico, apparently because of the great profusion of mountain-laurel that grew there and is still abundant in the area.

May 51

PARTRIDGE-BERRY

MOUNTAIN-LAUREL or CALICO-BUSH

SHEEP-LAUREL or LAMBKILL
Kalmia angustifolia Heath family

Mid- or late May to late June
6"- 36" high

A small shrub with nearly straight, ascending stems and branches. Small, narrow, thin but leathery leaves are elliptical to lance-shaped, dull olive-green above, often rusty spotted, lighter green underneath, persistent or evergreen into second year. Older, lower leaves tend to droop, while new, light green leaves stand nearly upright. Flowers very similar in shape and structure to those of mountain-laurel except these flowers are smaller and are deep crimson-pink. One other difference is that new growth and new leaves develop as extensions of the previous year's stems and branches before the plant blossoms, so that when the flowers do develop at the top of the previous year's growth, they now appear on the sides of the stems and sort of encircle the stem, below the new growth of the season.

Frequent to common in dry, or, more often, in moist, sandy areas and around the edges of bogs where it sometimes becomes abundant. Believed to be poisonous to sheep and cattle, hence, its name of lambkill.

SWAMP MAGNOLIA or SWEET BAY
Magnolia virginiana Magnolia family

Late May to mid- July
A tree, 8'- 30' high (in NJ)

This tall shrub or small tree has smooth, light gray bark on young stems that become darker gray and scaly on older branches and trunks. New twigs are green, rather slender, and covered with a downy coating. Buds are bright green and decidedly hairy. Leaves are simple, somewhat leathery, and blunt-tipped, 3"- 6" long, glossy green above and distinctly whitish underneath, spicy fragrant when crushed, and, in our area, usually persist throughout most of the winter. Flowers are creamy-white, large, from 2" to 4" across, globular or cup-shaped, very fragrant, and occur singly at the ends of branches. The fruit that develops in August is cone-like, oval, about 2" long, and contains a number of bright red, shiny, flesh-covered seeds that become suspended from the cone by thread-like cords.

Frequent to common in swamps and wet woods, particularly in mixed hardwood swamps and boggy wooded areas throughout the pine barrens. Although this small tree (which reaches heights of up to 70' in the South) clearly does not fit the definition of a wild flower, its large, white flowers are so beautiful and so fragrant, and the tree is so characteristic of pine barren swamps, that it simply had to be included in this book. When in bloom, the atmosphere in pine barren swamps can be laden with the heady fragrance of swamp magnolia.

SHEEP-LAUREL or LAMBKILL

SWAMP MAGNOLIA or SWEET BAY

COW-WHEAT
Melampyrum lineare Figwort family

Mid- or late May to end of August
4"- 18" high

 A small, inconspicuous, annual herb with a slender, wiry, grayish-green stem and, usually, with opposite, wide-spreading branches. Leaves in pairs and opposite, with lower leaves narrow, linear to lance-shaped, without teeth, on short stalks. Upper stem leaves usually with two to four bristle-pointed teeth, or lobes, at base of leaf. Frail, small, greenish-white flowers are tubular and cylindrical, opening into two lips, the upper lip arched, the lower lip three-lobed and tinged with straw-yellow at the tip. Flowers develop singly from between the leaves and are less than 1/2" long.

 Common, often becoming a weed, in partially shaded borders of dry woods and roadsides. Plants may be somewhat parasitic on roots of other vegetation.

BLADDERWORTS
Utricularia spp. Bladderwort family

Mid-May to late September

 Bladderworts are carnivorous herbs (see pitcher-plant description on page 48) that grow without roots, in either water or in wet, sandy soils. Instead of roots, modified leaves are located at the bases of their flowering stalks, either underwater (aquatic) or in mud (terrestrial), and are divided into fine, hair-like segments. Minute bladders located on these segments are capable of sucking in microscopic forms of animal life that are digested and their nutrients absorbed into the tissues of the plants. Flowers of bladderworts are two-lipped and are borne singly or in small, loose clusters at the top of their flowering stalks. There are a number of bladderwort species in the pine barrens. All but two have yellow flowers.

PIN-LIKE or
SLENDER BLADDERWORT
Utricularia subulata Bladderwort family

Mid- or late May to early September
1"- 6" high

 A terrestrial, or very shallow water, species with speck-sized bladders on underground branches. Leaves, if present, tiny, linear, half-buried in soil or wet sand. Flowering stem erect, wiry, very slender, and bears, sequentially, from one to 10, usually only three or four, small, 1/4", bright yellow flowers. Plant may be smallest terrestrial plant species in pine barrens, often only 1"- 2" high.

 Common in wet sand and sphagnum and along the edges of shallow, sandy, often ephemeral, ponds and pools.

COW-WHEAT

PIN-LIKE or
SLENDER BLADDERWORT

FIBROUS BLADDERWORT

Utricularia fibrosa Bladderwort family

Mid- or late May to early September

4"- 16" high

An aquatic species with mats of thread-like stems that float in shallow water. Flowering stems bear two to seven large, 3/4", yellow flowers, each on its own long stalk. Upper and lower lips of flowers broad, nearly equal. Spur narrow and short, only as long as, or barely longer than, lower lip (not illustrated).

Common in shallow, sandy ponds, pools, and edges of slow-flowing streams.

SWOLLEN BLADDERWORT

Utricularia inflata Bladderwort family

Mid-June to early September

Up to 10" high

A completely aquatic species with a circle of "floats" consisting of four to 10 air-filled arms radiating out on the surface of the water like the spokes of a rimless wheel. In addition, this plant has a mass of submerged stems on which there are many leaves, borne singly, and divided into thread-like segments that bear small bladders. From three to seven or more large, bright yellow flowers develop at or near the tips of the flowering stems that rise from the centers of the floating "wheels."

Frequent in small ponds and reservoirs throughout the pine barrens.

HORNED BLADDERWORT

Utricularia cornuta Bladderwort family

Mid- or late June to early August

2"- 12" high

One of the more conspicuous and beautiful of the several species of yellow bladderworts occurring in the pine barrens. This is a terrestrial species whose under-mud stems are very slender, delicate, thread-like, and so fine that they often are very difficult to find. Flowering stems bear from one to five bright yellow, two-lipped flowers, each with a conspicuous palate, and a prominent, downward projecting yellow spur that angles away from the lower lip. Flowers measure up to 3/4", with spurs from 5/16" to as much as 9/16".

Frequent to common in wet, sandy bogs and open, wet savannas.

May 57

SWOLLEN BLADDERWORT

HORNED BLADDERWORT

PURPLE BLADDERWORT
Utricularia purpurea Bladderwort family

Mid- July to early September
3"- 6" above water surface

This aquatic bladderwort has long stems that float beneath the surface of the water, bearing whorls of bladderwort-like leaves. The leaves are long-stemmed with many hair-like segments, and with bladders at the ends of some of the segments. Deep pink to purplish flowers are borne singly on slender stems, and the lower lip of each flower has two pouch-like lobes.

Occasional and local in ponds and slow-moving streams.

One of only two purple-flowering bladderworts found in the pine barrens. The other is described below.

REVERSED or RECLINED BLADDERWORT
Utricularia resupinata Bladderwort family

July and August
1"- 4" above water surface

The single flower of this species is tipped backward on its stem and faces upward so that its spur is almost horizontal.

This small terrestrial species occurs in very wet, mucky soils and in very shallow waters along the edges of ponds. Very rare in the pine barrens but is known to occur in the Wharton State Forest.

LARGE or LILY-LEAVED TWAYBLADE
Liparis liliifolia Orchid family

Late May to late June
4"- 10" high

This small, inconspicuous but attractive perennial orchid has two large, ovate, shiny, lily-like, light green leaves, 2"- 4" long, that develop from a small, roundish, thickened, underground stem, or bulb. Flowers are numerous and develop in a loose, elongated, upright cluster on a single stalk. The most conspicuous part of each individual flower is the broad, wedge-shaped, brown- or lavender-purple lower lip. Other flower parts very narrow, some almost thread-like.

Rare in moist woods and thickets. May have some preference for low habitats strewn with old bricks and mortar from early industries where lime was used in the construction of those facilities. This author knows of only one site for this flower that is within the pine barrens.

PURPLE BLADDERWORT

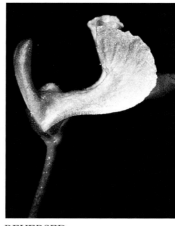
REVERSED or
RECLINED BLADDERWORT

D.H. ("Bart") Ulmer, MD

LARGE or LILY-LEAVED TWAYBLADE

FLY-POISON
Amianthium muscaetoxicum Lily family

Early to late June
1'- 4' high

A perennial herb with a thick bulb from which rise tall, narrow, green, grass-like, basal leaves that are somewhat blunt at their tips. One or more tall, smooth stalks, that may have a few bracts, also develop, and each of these will bear a dense head of small, white flowers, with each individual flower having six small, rounded petals. At first, the developing flower head is sort of cone-like but then becomes more cylindrical as it matures. Flowers are white at first, turning greenish or bronze-green with age. Superficially, the heads of these foam-like flowers may look somewhat like the flowering heads of turkey-beard but their habitats are much different.

Occasional and local in low, sandy, open woodlands, and along wooded edges of ponds and slow-moving streams. All parts of this plant are believed to be poisonous. The name *muscaetoxicum* comes from the Latin *musca*, for "fly", and *toxicum* for "toxic", thus, "poisonous to flies."

ROSE POGONIA or SNAKE MOUTH
Pogonia ophioglossoides Orchid family

Early June to early July
8"- 18" high

A slender perennial herb with fibrous roots that propagates itself by runners as well as by seed. Leaves usually two in number, with the larger, lower one midway up the smooth stem, the smaller, upper one (bract) just below the blossom. Sometimes a third, long-stemmed leaf develops from the root. The delicate, showy flowers on top of the stems range from pale lavender to crimson-pink, or to bright rose, with petals and sepals of equal length overhanging a beautiful, spatulate-shaped lower lip. This lower lip, called a "beard," bears three rows of fleshy hairs tipped with yellow and serves as a landing platform for insects. Usually, there is only a single flower on each plant, but occasionally, two-flowered specimens do occur.

Frequent in peaty bogs and damp, grassy areas, including roadsides. The name snake mouth or adder's mouth is for its bearded lip. This is the second of the three pink, spring orchids to bloom. This will be closely followed by grass-pink. The first was arethusa.

June 61

FLY-POISON

ROSE POGONIA or SNAKE MOUTH

GRASS-PINK
Calopogon tuberosus = pulchellus Orchid family

Early or mid-June to late July
6"- 24" high

From a round, underground stem or small bulb, this perennial herb sends up a smooth flowering scape, at the base of which there is a single (rarely two) linear, grass-like, green leaf. The flowering scape bears a loose cluster of from one or two to 10 or 12 individual flowers that open sequentially up the stem. Lateral petals and sepals, which are much alike, are pink, or rose-purple, or magenta-crimson, and seem to vary considerably in intensity of color from plant to plant. The upper petal, or lip, broadly triangular and dilated at the apex, is bearded with yellow-, orange-, and red-colored hairs. There is an important difference between grass-pink and both arethusa and rose pogonia in that in grass-pink, the bearded lip is the *uppermost* petal.

Frequent in peaty bogs, boggy meadows, and damp, grassy areas, including roadsides. This is the last of the three pink orchids that occur in quick succession each spring in suitable low, damp habitats.

CHAFFSEED
Schwalbea americana Figwort family

Early June to early July
1'- 3' high

A finely hairy, perennial herb with alternate, hairy, green leaves, with the lower leaves longer, ovate or lance-shaped and three-veined, upper leaves progressively smaller, gradually reducing to narrow bracts. Leaves attached directly to main stem, without stems of their own. Flowers are borne singly, on short stalks, in axils of upper leaves, all the flowers together forming a tall, loose spike of leafy-bracted blooms. When in full bloom, individual flowers are large, 1"- 1 1/2" long, hairy, and appear to be inflated. In reality, they are two-lipped, somewhat like a snapdragon, and are pale lemon-yellow and dark maroon in color.

Damp, sandy spots. Formerly more widespread in southeastern New Jersey. Now known from only one or two sites in the pine barrens, one of these along the shoulder of a well traveled county highway. An extremely rare and highly endangered pine barrens plant species.

June 63

GRASS-PINK

CHAFFSEED

PINE-BARREN SANDWORT

Arenaria caroliniana =
Minuartia caroliniana Pink family

Early June to late July
4"- 8" high

This perennial herb develops from a long, deep tap-root and is characteristic of the most bare stretches of white sand. Dense tufts of tiny, awl-shaped, overlapping, lower leaves are matted together so closely on the surface of the sand that they resemble a moss. From these mats, numerous, delicate, star-like, five-petaled flowers, white with greenish centers, are supported on slender, branching flower stalks. Upper parts of these stalks, just below the flowers, are somewhat glandular and sticky, so that tiny grains of sand and small insects, such as tiny flies, mosquitoes, and ants, are sometimes found stuck to these upper stalks.

Frequent, sometimes becoming abundant, on otherwise bare, white, dry, sandy stretches in the pine barrens. Sometimes called "long-root" for its long, deep tap-root.

WATER-SHIELD

Brasenia schreberi Water-lily family

Early June to mid- August
Leaves float on water surface
Flowers just above water surface

As do other water-lilies, the roots of water-shield, which are actually thickened stems, grow in mud under the water. The small, oval, shield-shaped leaves, 2"- 4" across, rise to the water surface on long stems to develop and float on the water. Two features distinguish these leaves: first, they are entire, without any V-shaped notch, and their stems are attached to the *centers* of their leaves; second, they are covered on their undersides with a thick, gelatinous coating. Small, 3/8"- 3/4" across, inconspicuous but attractive red-purple or maroon flowers rise just above the water surface and have six curled back sepals and petals and a dozen or more protruding stamens.

Occasional and local in ponds, sluggish and dammed-up streams, and other quiet waters.

PINE-BARREN SANDWORT

WATER-SHIELD

FRAGRANT WATER-LILY
Nymphaea odorata Water-lily family

Early June to late September
Leaves float on water surface
Flowers just above water surface

An aquatic, perennial herb with a thick, horizontal rootstalk down in the mud under the water. Leaves appear at the water surface on the ends of long, underwater stems and float on the water. The smooth, shining, dark green, almost circular or somewhat heart-shaped leaves are large, 4"- 8" in diameter, pinkish-purple underneath, and have a deep V-shaped notch in from the margin toward the supporting stem. The large, white, fragrant flowers, which develop on the tops of similar long stems, are 3"- 5" across and are made up of four green sepals and many rows of pointed, white or pinkish petals. Flowers open mornings but close by nightfall and on dark, cloudy days. Although normally almost pure white, sometimes the pink color usually present on the undersides of the petals will suffuse the entire flower with a pinkish tinge.

Common to abundant throughout the pine barrens. This is the familiar, fragrant, white water-lily seen on almost every small lake, pond, backwater, sluggish stream, cranberry bog reservoir, and feeder canal.

GOAT'S RUE
Tephrosia virginiana Pea or Bean family

Early June to early July
6"- 24" high

A perennial herb with a few to many erect stems growing in a dense cluster from a tough, stout, fibrous root. Stems and leaves silky with fine whitish hairs but, according to some authorities (Fernald 1950), the form found in the pine barrens tends to be smooth, var. *glabra*. Leaves all pinnately compound, that is they are divided into 9-25 small, finely toothed leaflets arranged side by side on each side of a common stem, plus one leaflet at the tip. Flowers are crowded in clusters at the tips of the branches, with each individual flower having a conspicuous, central, upper yellow petal, or standard (banner), while its lower keel and side (wing) petals are pink or pale purple.

Frequent in dry, shady, open woodlands and road edges.

June 67

FRAGRANT WATER-LILY

Doris Boyd

GOAT'S RUE

PRICKLY PEAR
Opuntia humifusa Cactus family

Early June to early July
6"- 18" high

A sprawling plant that develops from fibrous roots and sometimes forms large mats. Although generally prostrate, this plant may occasionally grow in somewhat upright positions. Its growth pattern consists of large, oval, thick, swollen, and fleshy green segments, or joints, rather tightly joined together, end to end. These segments usually do not have any spines, but they are armed with minute, barbed, hair-like bristles that are easily detached from the plant. When these are touched, their barbs enter one's skin, and, unless removed, can become embedded and cause painful inflammation. The large, showy, 2"- 3" flowers are yellow, sometimes with a reddish center. Each individual flower remains open for only a single day, after which it begins to turn into a pulpy, red or reddish-purple, oval fruit that, when ripe and peeled, may be eaten fresh or stewed.

Frequent in open, dry, sandy soils, especially in disturbed areas and along the outer fringes of the pine barrens.

SWAMP or CLAMMY AZALEA
Rhododendron viscosum Heath family

Early June to late July
3'- 8' high

A much branched, woody shrub with smooth, gray-brown bark and hairy twigs. Leaves ovate or lance-shaped and yellow-green but rather whitish underneath, with a few scattered hairs above and on the veins underneath. Flowers are white or rarely pinkish, very fragrant, and appear in terminal clusters *after* the leaves unfold. Outside surfaces of the flowering tube are glandular and are covered with ruddy, sticky hairs. Stamens are prominent and extend out beyond the flower's corolla, with the pinkish pistil being even longer than the stamens. Rarely, a shrub may be found where its flowers are deep pinkish rather than white. This may be confused with the earlier flowering pinxter flower, *R. nudiflorum*, but pinxter flower is rather rare in the pine barrens and, more important, blooms from early to late May *before* its leaves unfold.

Common in bogs, swamps, and wet woods. Swamp azaleas are very fragrant and seem to fill a gap between the flowering times of swamp magnolia and sweet pepperbush, with the three, together, providing an almost continual fragrance from late May to early September in pine barren swamps.

PRICKLY PEAR

SWAMP or CLAMMY AZALEA

NARROW-LEAVED SUNDROPS
Oenothera fruticosa = linearis Evening-primrose family

Early or mid-June to early August
1'- 3' high

A fibrous rooted perennial herb with a simple stem, or branched above, variously hairy, but nearly smooth beneath the flowers. Leaves slender, narrowly elliptical to lance-shaped or nearly linear, usually without teeth. Day blooming flowers open in early morning and remain so for only one single day. Flowers with four broad, bright yellow petals, indented at their mid-tip, with eight distinct orange stamens, and a prominent, cross-shaped stigma in the center. Seed pods long, narrow, twice as long as wide. and ribbed. Not to be confused with frostweed which has five rounded petals.

Frequent in dry, open ground throughout the pine barrens.

INDIAN PIPE
Monotropa uniflora Indian Pipe family

Early June to early September
3"- 10" high

The most striking feature of this unusual plant is that, because the plant lacks any chlorophyll, it has no ability to manufacture its own food. Instead, it must get all its nourishment from partially decayed organic material in the soil, which it obtains by means of a mass of matted, brittle roots. Thus, this is known as a saprophytic plant. From the roots of this plant, one or two, or, more frequently, a cluster of thick, translucent, waxy-looking, white or pinkish-white plant stems develop, each without leaves except for a few scaly bracts. Each stem bears a solitary (rarely two) nodding flower, which hangs with its open end down and which has sepals, petals, stamens, and other flower parts. After flowering, the seed capsule that develops from the flower unbends from its nodding flower position and becomes erect. The entire plant now becomes black and brittle and stands like a silent sentinel in the woods, giving rise to its other names of "corpse-plant" or "ghost-flower."

Occasional to frequent in dry or slightly damp woodlands, especially in areas with a reasonable amount of humus.

NARROW-LEAVED SUNDROPS

INDIAN PIPE

ORANGE MILKWORT
Polygala lutea Milkwort family

Early June to mid-October
6"- 16" high

A biennial or perennial herb with smooth, erect stems that develop from fibrous roots. Stems often cluster together and may be branched. Basal leaves oblong, broader toward outer tip, and blunt, often forming rosettes at ground level. Stem leaves alternate, lance-shaped, often with broadest part toward outer tip, smaller than basal leaves. Flowers in terminal, dense, blunt, almost spherical, clover-like heads of numerous bright orange-yellow blossoms, with each head usually on top of a long, leafless stem. This is such a showy plant, it simply cannot be overlooked.

Common in wet, sandy soils, bogs, and damp, grassy roadsides.

YELLOW LOOSESTRIFE or SWAMP CANDLES
Lysimachia terrestris Primrose family

Mid-June to mid-July
1'- 2 1/2' high

A perennial herb with erect, simple or sparingly branched, smooth stems that develop from long, creeping, underground stems. Leaves usually paired and opposite, lance-shaped, pointed at both ends, and closely attached to stems. A distinctly characteristic, tall, slender, terminal flowering spike bears small, star-like, yellow flowers with five petals that usually are spotted with reddish dots or streaks. Flowers are attached to main stem with long, slender stalks of their own. In late summer and early autumn, small, jointed bulblets develop out of the axils of the leaves, giving rise to this sometimes being called "bulb-bearing" loosestrife.

Common in open swamps, meadows, marshes, and bogs. May become a nuisance weed in commercially operated cranberry bogs, sometimes becoming so abundant they provide an overall misty, yellow color to the bogs.

ORANGE MILKWORT

YELLOW LOOSESTRIFE or SWAMP CANDLES

HYSSOP SKULLCAP
Scutellaria integrifolia Mint family

Mid-June to mid-July
6"- 24" high

A perennial herb that develops from creeping, underground stems. Stems above ground finely to densely hairy and bear leaves that are lance-shaped, without marginal teeth or without being notched, and without stems of their own. Some lower leaves may have toothed blades on their stalks, but these leaves soon fall off. Flowers blue or bright, light violet, often whitish underneath, arranged in one or more loose, terminal clusters. Petals of each flower collectively form a tube of about 1" that curves upward, ending in two lips: the upper lip arched and hood-like, the lower flat and slightly flaring, with the two side lobes seemingly connected with the upper, rather than lower, lip.

May occur occasionally as an introduced species in dry or moist ground and along the edges of fields and woods in pine barrens.

LARGE or AMERICAN CRANBERRY
Vaccinium macrocarpon Heath family

Mid-June to mid July
Trailing stems up to several feet long
Branches (uprights) 6"- 10" high

A low, trailing, somewhat woody, perennial sub-shrub, or vine, with very long, slender, lithe stems that trail along the ground and send up shorter, upright branches, the whole forming dense ground mats of thickly interwoven stems and branches. Leaves small, thick, leathery, evergreen, paler underneath than above, elliptical or oblong, rounded at the tip, not over 5/8" long, usually less, and very closely attached to main stems or branches. From two to six, usually two to four, white-pinkish, nodding flowers are borne in small, open, loose clusters on top of erect, 1/2"- 1" stems (uprights). Long-stemmed, individual flowers are deeply four-cleft, with the four lobes strongly curled back, exposing the elongated, orange-yellow stamens that are compressed into a narrow cone. Fruits yellow-green, ripening in the fall to bright red berries.

Cranberry is a native, wild, North American plant that is common in wet, sandy, peaty bogs, low grassy areas, and along the edges of most cedar streams in the pine barrens. This is the cranberry that is cultivated and harvested commercially. It is reported that the form of the flower gave this plant its name. Early colonial settlers imagined it resembled the neck (stem), head (curled back lobes), and beak (stamens) of a crane, a large European bird, so they called it "crane-berry."

June 75

HYSSOP SKULLCAP

LARGE or AMERICAN CRANBERRY

COLIC-ROOT
Aletris farinosa Lily family

Mid-June to late July
1'- 3' high

A perennial herb with a thick, spreading, basal rosette of narrow, lance-shaped, long-pointed, yellow-green leaves growing from a short, tough, thick underground stem. Leaves up to about 8" long and 1/2" wide. The flowering stalk rises in the middle of these leaves, bearing a few bract-like leaves, and a stiff, terminal spike, 4"- 12" long, of small, short-stemmed, tubular, white flowers. All six sepals and petals are joined together to form the tubular flower which is up to 1/2" long, six-toothed at the rim, and roughened on the outside with numerous short, scale-like points, giving it a granular appearance.

Occasional to frequent in dry, sandy soils in open woodlands, and along similar roadsides.

BOG-ASPHODEL
Narthecium americanum Lily family

Mid-June to late July
10"- 30" high

A perennial herb from a stout rootstalk, with basal, narrow, linear, green, iris-like leaves and an erect stalk bearing a terminal spike of yellow flowers. The short, stiff, and erect basal leaves, 4"- 8" high, grow up like short grass from the bases of the flowering stalks, but the leaves up on the flowering stalks are few and bract-like. The 2 1/2"- 3" flowering head bears a dense cluster of small, six-petaled (actually three petals and three sepals) bright, yellow, star-like flowers, each with six yellow stamens. Where this plant occasionally grows in some abundance and in great patches, the total panorama of yellow flower heads makes a golden sheen over a bog that can be seen for quite a distance. After blooming, the spent flower heads develop into attractive, spindle-shaped, reddish-brown seed capsules.

Confined to wet, sandy bogs and open, boggy savannas in the very heart of the pines. Rare and local in the pine barrens.

June 77

COLIC-ROOT

BOG-ASPHODEL

NARROW-LEAVED WHITE-TOPPED ASTER   Mid-June to mid-August
Aster solidagineus = Seriocarpus linifolius   1'- 2 1/2' high
Aster or Composite family

Stiff, sturdy, basically smooth stems arise, either obliquely or erect, from a stout, perennial root. Stems branch upward to a flat-topped inflorescence. Leaves without teeth, linear or narrowly oblong, as much as five times as long as wide, or 1"- 3" long. Upper leaves attach directly to stem, lower leaves have own stalks. Flowering heads in flat-topped clusters of small, aster-like flowers. Individual flowers with few, only four to six, short, white or cream-colored rays. Disk flowers cream-colored. Bracts of flower heads whitish, with green tips (not illustrated).

Frequent in dry, sandy woods and open roadsides.

TOOTHED WHITE-TOPPED ASTER   Mid- or late June to late August
Aster paternus = Seriocarpus asteroides   1'- 2' high
Aster or Composite family

This white-topped aster differs from the narrow-leaved form described above mainly in that there may be a basal rosette of sparsely toothed, egg-shaped leaves, and its upper, narrow-based, spatula-shaped leaves have several teeth beyond the middle.

Occasional in dry woods and clearings. Not as frequent as narrow-leaved white-topped aster. Begins to bloom about one week later than the narrow-leaved species.

PENCIL-FLOWER   Mid-June to mid-September
Stylosanthes biflora Pea or Bean family   6"- 24" high

A perennial herb with wiry, branching stems that bear narrowly lance-shaped or elliptical, slightly hairy, and bristle-tipped trifoliate leaf segments ("clover-leaf" arrangement), with the central or end segment having a longer stem. Small, orange-yellow, pea-blossom shaped flowers develop singly or in small clusters at the ends of the branches.

Occasional in open, dry, sandy ground such as edges of woods and roadsides.

TOOTHED WHITE-TOPPED ASTER

PENCIL-FLOWER

ARROW-HEAD
Sagittaria engelmanniana Water-plantain family

Mid-June to mid-September
8"- 24" high

A perennial, aquatic herb, with a thickened base, that grows with fibrous roots from a creeping undermud stem. Leaves long-stemmed and distinctly arrow-shaped, with very long, narrow, forward-projecting blade, and two very narrow, backward-projecting lobes, each one-quarter to one-half the length of the blade, but scarcely more than 1/8" in width. Overall leaf may measure up to 8" or 10" in length and from 1"- 4" in width. Flowers are borne on short stems, in whorls of three, near the summit of smooth, flowering stalks. Each flower, approximately 1/2"- 1" wide, has three green sepals, three white petals, and many stamens, each petal measuring from 1/3"- 1/2" in length.

There are many species of arrow-heads, and there is considerable variation between, and even within, the species. For positive identification of this species, it may be necessary to count the number of stamens, which should be between 15 and 25. If less or more, consult a botanical text for other possible species.

Common to abundant in shallow, fresh water bogs and ditches. This is the common arrow-head of the pine barrens. The stout, underwater stems of this and other species of arrow-head have a very high starch content so they are a good source of food for many types of waterfowl. In earlier times, these were a staple food for Native Americans, and early colonial settlers called them "duck"- or "swan"- potatoes.

FALSE or VISCID ASPHODEL
Tofieldia racemosa Lily family

Late June to mid-July
6"- 24" high

A small, perennial herb with a basal tuft of erect, grass- or iris-like leaves, 2"- 6" high, and a flowering scape, both of which grow from a thickened rootstalk. The flowering scape bears a small bract midway up the stem, and the scape is covered with fine, short hairs that are slightly sticky (viscid). Atop this scape is a loose, open head of small, white or greenish-white, star-like, six- petaled flowers that are mostly clustered in threes and fours. When Witmer Stone (1910) first saw these flowers in 1904, he referred to "the white spikes of starry flowers" as being "like miniature turkey-beard."

Rare. Found only in wet, peaty sands and open cedar bogs and savannas in the very heart of the pine barrens.

ARROW-HEAD

FALSE or VISCID ASPHODEL

RAGGED FRINGED or GREEN FRINGED ORCHID

Late June to very early July
1'- 2 1/2' high

Habenaria lacera = Platanthera lacera Orchid family

From a cluster of thick roots, a smooth, slender stem arises to bear several narrow, lance-shaped leaves and a loose spike of 5 to 15 translucent white, green, or greenish-yellow flowers. Lower leaves may be up to 8" long, upper leaves decrease in size up the stem. The two most conspicuous parts of the flowers are the lower lip and the spur. The lower lip, actually a petal, may be 1/2" or more long and is deeply cut into three parts, with each part then deeply cut into a thread-like fringe so finely divided that the entire lip looks very ragged or tattered. The spur that drops down behind the flower may be as much as 3/4"- 1" in length, is curved, and is slightly thickened at the end.

Uncommon to occasional in open, wet, sunny places such as meadows, marshes, and open wet thickets and woods. Not supposed to be in the pine barrens, but this photograph was taken in a small, grassy opening in a low woodland near the old village of Retreat, Southampton Township, Burlington County, well within the pine barrens.

SPOTTED or STRIPED WINTERGREEN

Mid- or late June to mid-July
1'- 2 1/2' high

Chimaphila maculata Shinleaf family

A small, perennial, semi-shrub that arises from creeping, underground stems. Smooth, erect stems bear whorls of thick, dark green, evergreen leaves that are conspicuously striped, not spotted, with white or greenish-white along the veins. Leaves are entire, lance-shaped, taper to a sharp point, and have sharp, pointed teeth at rather wide intervals along their margins. Several waxy-appearing white flowers, each with five sepals and five rounded petals and a dome-shaped green center (ovary) hang facedown from the ends of their stems, which spread outward from the main stem, the whole forming a small, loose spike.

Frequent in dry, upland, oak and pine woods throughout the pine barrens.

RAGGED FRINGED or
GREEN FRINGED ORCHID

SPOTTED or STRIPED WINTERGREEN

PIPSISSEWA or PRINCE'S PINE

Chimaphila umbellata Shinleaf family

Mid- or late June to mid-July
6"- 12" high

Rising from creeping, trailing, slightly woody, underground stems, the erect stems of this perennial semi-shrub bear whorls of narrowly wedge-shaped, shining bright green, sharply toothed, evergreen leaves that are broad and bluntly pointed at the tip and become narrower as they taper toward the base. Several waxy-appearing, white or pale pinkish flowers are produced in a loose cluster above the leaves. Individual flowers have five concave, roundish petals that turn backward as the flower matures, and at their base, next to the dome-shaped greenish center, is a circle of pale magenta.

Dry woods but quite uncommon in the pine barrens. This photograph taken in a disturbed area along the edge of Lebanon State Forest.

MILKWEEDS

Asclepias spp. Milkweed family

Mid-June to early August

Milkweeds are tall, perennial herbs whose most distinguishing features include the thick, white sap that oozes from any cut or broken surface of the plant (except in butterfly-weed), the clustered arrangement of their flowers, the unique structure of their individual flowers, and their long seed pods containing seeds attached to white tufts of floss.

A typical milkweed flower is five-parted and, when in full bloom, its five, sepal-like corolla segments are turned far backwards down its stalk. Above these are five upright hoods (the corona) that comprise the main part of the flower. Inside each hood is a curved horn that may or may not project up above the hood. This non-technical description applies to all the milkweed descriptions in this book.

PIPSISSEWA or PRINCE'S PINE

BLUNT-LEAVED MILKWEED

Mid-June to mid-July

Asclepias amplexicaulis Milkweed family

2'- 3' high

A stout, smooth, perennial herb with erect stems that sometimes start curved, or almost prostrate, near their base but then become erect. Leaves opposite, oblong, blunt at both ends, somewhat heart-shaped at base, and clasp stem. Margins of leaves *distinctly wavy*. Flowers of blunt-leaved milkweed develop in a single (rarely two) terminal cluster and are greenish, stained with magenta-purple, or pinkish, with the horns protruding above the pinkish hoods. Seed pods are long, slender, and nearly smooth.

Frequent in open, dry, sandy ground, edges of woods, and roadsides. This is the first of several milkweeds to bloom in the pine barrens.

COMMON MILKWEED

Mid- or late June to late July

Asclepias syriaca Milkweed family

2'- 4' high

A tall, coarse, perennial herb with stout, rarely branching stems that are fairly hairy above. Oblong to oval, paired leaves, on short stems, are finely but densely downy beneath, smooth above when mature, pointed or blunt at tips, rounded or slightly heart-shaped at base. Veins join midrib almost at right angles, curving only slightly as they approach the margin. Flowers numerous, in one or several clusters on long stalks from upper axils of the leaves. Flowers vary from greenish-purple, or pale purple, to crimson-pink or pink-lilac. Hoods 1/8"- 1/6" long. Horns curved, not surpassing hoods, but easily seen. Seed pods erect on curved stalks, hairy, ribbed, and with a warty surface.

Common to abundant as a weed in old cultivated fields and in waste and other disturbed areas, including roadsides. Not a native pine barrens species, into which it undoubtedly has been introduced.

BLUNT-LEAVED MILKWEED

COMMON MILKWEED

RED MILKWEED

Asclepias rubra Milkweed family

Mid- or late June to late July
1'- 4' high

A tall, smooth, perennial herb with paired, broadly lance- or egg-shaped leaves that are rounded at the base and taper to long, pointed tips. Leaves attached directly to main stem, or nearly so. Leaf veins almost form right angles with the midrib, curving only slightly as they approach the margin (compare with swamp milkweed, page 90). Flowers develop in small, terminal clusters of purplish-red flowers with tall, thin hoods about 1/3" long and with straight horns that are almost as long as the hoods. Seed pods are smooth.

Formerly frequent in pine barren bogs and considered to be the typical milkweed of the pine barrens, but now seems to have become rather rare.

BUTTERFLY-WEED

Asclepias tuberosa Milkweed family

Mid- or late June to early August
1'- 2 1/2' high

Stout, hairy, and erect or sometimes oblique, frequently branching stems of this perennial herb develop from a deep, stout, tough, underground root. Alternate olive-green, roughly hairy leaves are long, narrow, lance-shaped, and pointed at the tip, rounded or heart-shaped at the base, and are attached to the main stem with a very short stalk, or none at all. Butterfly-weed has an almost clear, or only slightly milky, plant juice rather than the thick, white, milky sap that is so characteristic of other milkweeds. Bright orange or orange-yellow flower heads develop at the tops of the branches as loose, flat-topped clusters of numerous individual, intricately shaped flowers. Seed pods are long, slender, spindle-shaped, and finely hairy, are borne on short stalks with an S-curve, and, as in other milkweeds, the seeds themselves are tipped with long, silky hairs.

Occasional in open, dry, sandy ground, especially in disturbed areas and along roadsides. Not considered to be a true pine barrens species, having apparently been introduced.

June 89

RED MILKWEED

BUTTERFLY-WEED

SWAMP MILKWEED
Asclepias incarnata Milkweed family

June to August
2'- 4' high

A tall, smooth, perennial herb that may have two downy lines on the upper part of its main stem and on the branches of the flower stalks. Leaves are opposite, narrow, lance-shaped, short-stemmed, and taper to the tip. Leaf veins form acute angles with the midrib, slanting toward the margins (compare with red milkweed). The small flowers of swamp milkweed develop in flattish clusters of flowers that vary from pink to rose-crimson, with hoods only about 1/8" long, and horns that are noticeably longer than the hoods. Slender seed pods taper to both ends.

Swamp milkweed is not supposed to be present in the pine barrens but there are several low, damp, roadside ditches where this has become established, apparently as introductions.

TEABERRY, CHECKERBERRY, or WINTERGREEN
Gaultheria procumbens Heath family

Mid- or late June to early August
2"- 6" high

A low, perennial, semi-woody sub-shrub with creeping, subterranean stems, from which arise erect, buff-brown or ruddy stems that bear small clusters of oval, thick, leathery, dark green, evergreen leaves. Leaves are shiny above, pale beneath, with slightly rolled under margins and a few bristle-tipped teeth, or with smooth margins. New, young leaves that appear early in May are yellowish to reddish-green. The usually solitary flowers are waxy-white or slightly pinkish and hang on downward-curving stems out of the axils of the leaves. Individual flowers are urn- or bell-shaped and consist of five petals joined together to form a small bell beneath the leaves. By autumn, a dry, aromatic, deep red berry will have developed, and this will mature over winter and become more succulent in the spring.

This characteristic plant of the pine barrens is common to abundant in both dry and moist wooded habitats. The spicy, aromatic flavor of wintergreen is present in all parts of the plant: leaves, flowers, and berries. This is the active element in oil of wintergreen, which has been widely used as a flavoring for chewing gum and candies, as a scent for soaps, and as a camouflage for bad-tasting medicines.

SWAMP MILKWEED

TEABERRY, CHECKERBERRY, or WINTERGREEN

SUNDEWS
Drosera spp. Sundew family

Mid-June to late August

Sundews are herbs that are remarkable for their thread-like, spoon-shaped, or round leaves which bear numerous fine hairs tipped with sticky glands. Small insects that are attracted to these leaves are caught by their sticky hairs, which then bend around toward the captive and envelop it, thus completing its entanglement. A digestive matter is then exuded by the glands and this enables the plant to actually absorb nutrients from the insect's body into the tissues of the plant. Such "insectivorous" or "carnivorous" plants are in this way able to supplement their normal intake of nutrition from the soil. (See pitcher-plant description on page 48.)

Flowers of sundews develop in a loose spike that is curved over at the top and bears a few to several flowers that open one at a time from the bottom to the top of the flowering stalk.

There are three species of sundews in the pine barrens. All are found in bogs, marshes, and other wet, sandy places.

SPATULATE-LEAVED SUNDEW
Drosera intermedia Sundew family

Early July to late August

4"- 9" high

This carnivorous herb has long-stalked leaves that are covered with short, glandular hairs that exude dew-like drops of a sticky nature to capture insects, but these leaves are sort of spoon-shaped, are blunt-tipped, and are two or three times longer than wide. This species is apt to be out in open sunlight, and the hairs on its leaves are often quite reddish in color. Flowers of this sundew are white and are borne in a similar manner to the thread-leaved sundew (see page 93), but usually the flowers are fewer in number, more like four to six, and appear one at a time on a short flowering spike. Flowers are not conspicuous and are often overlooked (not illustrated).

Common to abundant in and around open bogs, shallow ponds, and other open, damp, sandy locations. May be the most abundant of the three sundews in the pine barrens.

THREAD-LEAVED SUNDEW
Drosera filiformis Sundew family

Mid- or late June to late August
6"- 16" high

This is the most conspicuous, attractive, and first to bloom of the three sundews. From fibrous roots, leaves of this carnivorous herb develop as erect, unbranched stems that uncoil at their tips as they grow. These stems are covered with short, usually reddish, glandular hairs, the tips of which are covered with tiny, dew-like drops of a sticky liquid that enables the plant to capture small insects. From 4-16 rose-pink to ruby flowers develop in loose spikes on separate stalks that curve over at the top, with flower buds lined down one side. The lowest opens first, for one or two days, then closes, and the others follow in nearly daily succession up to the tip of the spike. Flowers open only in sunshine and usually only during morning hours, closing by noon or shortly thereafter.

Common to locally abundant in open, damp, sandy areas. A stand of these growing in a bog can be an impressive sight as the sun shines through behind the plants and their hair-tip secretions glisten in the sunlight.

THREAD-LEAVED SUNDEW

ROUND-LEAVED SUNDEW
Drosera rotundifolia Sundew family

Early July to late August
2"- 8" high

Like the spatulate-leaved sundew, this species also has long-stemmed leaves, but these leaves are distinctly round or circular in shape, usually broader than long, lie close to or almost upon the ground, and are covered with fine hairs that are likely to be more greenish than reddish. This lack of reddish coloration may be due to the fact that this species, at least in the pine barrens, is often more characteristic of shady cedar bogs and swamps, where it gets very little sunlight. This round-leaved sundew also bears white flowers that are few in number, usually in the range of four to six, are not very conspicuous, and are often overlooked (not illustrated).

Frequent to locally common in cedar bogs and swamps, where it often grows deep down in masses of sphagnum moss. May be the least abundant of the three sundews in the pine barrens.

PICKEREL-WEED
Pontederia cordata Water-hyacinth family

Mid- or late June to mid-September
1'- 4' high

A perennial, aquatic herb that develops from a thick, horizontal rootstalk embedded in mud under the water. Often grows in large patches. Leaves large, thick, glossy, deep green, broadly arrow- or heart-shaped, indented at the base producing two lobes, and tapered upward to a point (cordate), but these leaves are subject to great variation in size and shape. Leaves borne on long stalks, usually rising above the surface of the water, and range from 2"- 10" long and 1"- 6" wide. Flowering stalks are erect and the flowers are in dense, showy, 2"- 6" spikes, 1' or more above the water. Individual flowers are two-lipped, with each lip three-lobed. All the sepals and petals, collectively called the perianth, are deep violet-blue, often dotted with whitish hairs and yellowish-green spots. Individual flowers are very short-lived. Just below the flowering spike is a small, solitary, leaf-like bract (spathe).

Common in shallow ponds, sluggish streams, and wet ditches.

PICKEREL-WEED

SPREADING POGONIA
Cleistes divaricata Orchid family

Late June to early July
1'- 2' high

From a cluster of fleshy roots, a single, tall stem develops that bears a single flower. About midway up the stem, there is a single, clasping, oblong to lance-shaped leaf from 3"- 6" long, and there is a much smaller, bract-like leaf just below the flower. The flower itself is tubular in shape and is formed by having the petals and the narrow, trough-shaped lip joined together to form a slender tube, up to 2" long, which may be pale magenta to pink or almost white. Three conspicuous, long, up to 2 1/2", narrow sepals are dull greenish-brown and are widely spreading above the tubular flower.

Very local in moist, streamside, sandy pine woods and edges of bogs. Currently known from only one station in the pine barrens where it was recently "rediscovered."

GOLD-CREST
Lophiola aurea Bloodwort family

Late June to late July
1'- 2' high

A perennial herb with a slender stem that rises from a slender rootstalk. Stems are thinly hairy below, becoming densely covered with white, woolly hairs above. Leaves linear, narrow, and nearly smooth; basal leaves up to 1' high, upper leaves smaller, almost bract-like. Flowering cluster much branched and whitened with a dense coating of soft, matted, woolly hairs that surround each of the individual golden-yellow flowers. Witmer Stone, in *The Plants of Southern New Jersey*, 1910, best described this, writing "The dense, wooly covering of the flowers recalls the *Eidelweiss* of the Swiss mountains, and from the downy, white clusters the little yellow flowers peep out like tiny stars".

Local in bogs, cedar bogs, and open, boggy savannas, but where it does occur, it can sometimes be almost abundant. A characteristic species usually found only within the very heart of the pine barrens.

SPREADING POGONIA

GOLD-CREST

LADIES' TRESSES
Spiranthes spp. Orchid family

Late June to mid-October

Ladies' tresses are delicate herbs with small, white or greenish-white flowers that develop in a narrow spike from a cluster of thick roots, or a tuber. In most species, leaves develop at the base of the flowering stalk, though there may be a few smaller leaves up the spike. In some species, all leaves may have disappeared by flowering time. In some species, flowers spiral around the stalk. Individual flowers are small, and the side petals are joined to the uppermost sepal to form a sort of hood. In most species, the lower central lip flares a bit at the end, has crisped edges, and glistens (use hand lens).

Reference to a standard, technical, botanical manual is recommended for accurate species identification.

SPRING LADIES' TRESSES
Spiranthes vernalis Orchid family

Late June to early August
8"- 30" high

Spring ladies' tresses have grass-like leaves that remain present during the flowering period. A single row of 3- 15 white or yellowish-white flowers spiral up the stalk. Bases of flowers may be covered with a fine down. This species begins to bloom nearly three weeks before the next, or little ladies' tresses, does.

Found in open, sandy, moist or dry soils.

LITTLE LADIES' TRESSES
Spiranthes tuberosa = beckii Orchid family

Mid-July to mid-September
6"- 12" high

A very slender herb with ovate or egg-shaped basal leaves, which usually have disappeared by flowering time. Root a simple tuber. Flowers white, very small, usually in a close spiral up spike.

Occasional in dry fields and along roadsides.

SPRING LADIES' TRESSES

LITTLE LADIES' TRESSES

GRASS-LEAVED LADIES' TRESSES
Spiranthes praecox Orchid family

Early August to late September
8"- 30" high

Leaves very narrow, grass-like to even thread-like, up to 10" high. Flowers may be in a spiral arrangement or may simply be in a single line up one side of the flowering stalk. Flowers are white, sometimes veined with green on the lip (not illustrated).

This and the earlier spring ladies' tresses are our two taller species, but this just begins to bloom at about the same time the earlier species ends its blossoming period, so the two species just barely overlap.

Occasional to possibly frequent in wet pine barren swamps, bogs, meadows.

NODDING LADIES' TRESSES
Spiranthes cernua Orchid family

Early September to mid-October
6"- 18" high

A robust, late-flowering species, with flowers that may curve or arch slightly downward, thus "nodding." Leaves mostly at base, rather narrow, grass-like, becoming somewhat broader at the tip. Flowers white, whitish, creamy, or greenish-white in two or more spiral rows up their stalk.

Frequent in open, moist, grassy meadows, swamps, bogs, thickets, and along roadsides. The last and possibly the most common ladies' tresses to bloom in the pine barrens.

HELLEBORINE
Epipactis helleborine Orchid family

Late June to late August
1'- 2 1/2' high

A leafy-stemmed orchid with leaves that are strongly veined and vary considerably in size and shape, lower ones larger, up to 7" long, upper ones progressively smaller up stem. All leaves clasp stem, but lower leaves may have a short, narrow, stalk-like portion. Flowers in a tall, often one-sided, spike; each flower in the axil of a long, narrow bract. Sepals and lateral petals about 1/2" long, and green, usually suffused with madder, rose, or purple. Lip heart-shaped, forming a sac, with a triangular tip that is turned, or bent back under, the sac.

Not a pine barrens species but found in a disturbed area on the edge of Lebanon State Forest, where this photograph was taken. Introduced from Old World and becoming widespread in northeastern United States.

June 101

NODDING LADIES' TRESSES

HELLEBORINE

TURK'S-CAP LILY
Lilium superbum Lily family

Early July to mid-August
3'- 7' high

An absolutely superb, tall, showy lily that develops from a scaly bulb. The main stem bears numerous smooth, light green leaves, the lower ones arranged more or less in circles, upper ones mostly alternating. From the top of this stem, one or more nodding flowers are suspended from long, erect, or oblique pedicels. These large, showy flowers are remarkable for their completely reflexed sepals and petals (the perianth) that curl back so far their tips nearly touch, thus forming "turk's caps." As a result, the pistil, and the stamens tipped by their brown anthers, are completely exposed. The bright orange-red flowers are heavily freckled with brown or purple and each has a clearly defined greenish star at its center.

Occasional in wet meadows and damp roadsides throughout southern New Jersey, including the pine barrens. Stone (*The Plants of Southern New Jersey, 1910*) referred to this as *the* lily of southern New Jersey.

MEADOW-BEAUTY
Rhexia virginica Meadow-beauty family

Early July to mid- or late August
6"- 24" high

This perennial herb develops from fibrous roots that are distinguished by having small tubers, without runners (see Maryland meadow-beauty, page 104). The stout, smooth or slightly hairy stem is somewhat squarish, with four thin, lengthwise ridges or wings. Paired leaves are more or less oval, without stems of their own, and are conspicuously three- (usually) to five-veined. A few to several vivid pink or rose-colored to purplish-magenta flowers develop in profuse terminal clusters, each blossom with four broadly egg-shaped, or lop-sided, petals, and eight conspicuous, bright yellow stamens, the pollen-bearing anthers of which are long, narrow, curved, and appear to be set on their stems at right angles. Seed capsules are small, vase-like, broad at the base and taper to a long, narrow neck, which led Thoreau to call these "perfect little cream pitchers."

Common in moist, sandy meadows, open woods, and roadside ditches.

TURK'S-CAP LILY

MEADOW-BEAUTY

MARYLAND MEADOW-BEAUTY
Rhexia mariana Meadow-beauty family

Early July to mid- or late August
6"- 24" high

Superficially, a pale pink to light magenta version of the preceding, but with these differences: a slender, *round* stem, rather *hairy*, that grows from slender, horizontal *runners*. Leaves more narrow, elliptical, and spreading. Less common in the same general, but somewhat drier, habitats.

SLENDER YELLOW-EYED GRASS
Xyris torta Yellow-eyed grass family

Early July to late August
6"- 24" high

A perennial herb with flat, narrow, grass-like leaves and with flowering stalks considerably taller, usually, than the leaves. Base of plant enlarged, somewhat like a small bulb. Leaves very narrow, under 1/4", are often spirally twisted, and enclose flower stalk near base. Each of the compact, scale-covered, brown, cone-like flowering heads produces a number, but only one at a time, of bright yellow, three-petaled flowers that last only a short while. Flowers appear from the axils of the bracts that make up the flower head, usually opening in the early morning.

Common in wet, acid bogs, wet pine barrens, and sandy swamps.

CAROLINA YELLOW-EYED GRASS
Xyris caroliniana Yellow-eyed grass family

Mid-July to early September
3"- 18" high

A small, tufted, grass-like perennial with numerous fibrous roots that is similar to the slender yellow-eyed grass, but with the base of the plant soft and flat, not bulb-like. Leaves flat, linear, usually shorter than flower stalk, and enclose flower stalk at base. Leaves narrow, under 1/4", not usually twisted, but may twist as they grow older. Flower stalk bears, at summit, a dense, ovoid, blunt head of overlapping bracts or scales. Flowers bright yellow and three-petaled, as above, and appear one at a time from the axils of the scaly bracts (not illustrated).

Frequent to common in wet, sandy, peaty bogs, meadows, and swamps.

MARYLAND MEADOW-BEAUTY

SLENDER YELLOW-EYED GRASS

RED-ROOT
Lachnanthes caroliniana = tinctoria Bloodwort family

Early July to late August
8"- 30" high

A perennial herb with erect, flat, narrow, linear, pointed, basal leaves and with a diagnostic red plant juice that is most obvious in its roots. Flower stems are stout and erect, nearly smooth below but woolly-hairy above. Leaves up stem smaller than basal, or almost bract-like. Compact flowering head may be flat or rounded and is composed of multiple, densely woolly, dingy-looking, yellow flowers crowded together in a dense, branching cluster at the top of the stem.

Common in sandy swamps and bogs, becoming so abundant in cranberry bogs as to become an obnoxious weed.

LANCE-LEAVED CENTAURY
Sabatia difformis Gentian family

Early July to late August
1'- 3' high

A perennial herb with one or more slender, somewhat four-sided, stems that develop as small clusters from stout, underground stems. Leaves light green, egg- or lance-shaped, taper to a sharp, pointed tip, and have three to five ribs. A basal rosette of leaves has usually disappeared by flowering time. Flowering stalk branched near top, producing numerous five-petaled, creamy-white flowers that turn distinctly yellowish upon fading.

Common in swamps. bogs, and wet pine barrens and as a weed in cranberry bogs.

RED-ROOT

LANCE-LEAVED CENTAURY

PINESAP
Monotropa hypopithys Indian Pipe family

Early July to early September
4"- 12" high

A somewhat similar plant to indian pipe (see page 70). Again, this plant lacks chlorophyll, cannot manufacture its own food, and must get its nourishment from partially decayed organic matter in the soil, making it a saprophytic plant. Other similarities to indian pipe are that the stems of pinesap are often in clusters, appear to be translucent and waxy looking, and bear numerous bracts. Major differences from indian pipe are that the stems of pinesap are pale tan, yellowish, or reddish (indian pipe are white), and there are several, from 3-10, drooping, dull yellow to light crimson-reddish (not white) flowers in a small cluster at the top of each stem (only a single flower at top of indian pipe stem). After flowering, as in indian pipe, the fleshy, vase-shaped seed-pods become erect and black.

Occasional in dry woods in the outer fringes of the pine barrens. Most often found above the roots of oaks, pines, and beech.

HORSE-MINT
Monarda punctata Mint family

Early July to early October
1'- 3' high

An erect, perennial herb with pale gray, downy stems and opposite, narrowly oblong or lance-shaped, stalked leaves. Like most mints, this plant is somewhat aromatic and its stems are distinctly squarish or four-sided. Flowers develop in dense clusters, or rosettes, at the tops of the main stems and also at the ends of branching stems that develop out of the axils of principal stem leaves. Individual flowers are long, narrow, and markedly two-lipped, the upper lip continuing as a tube, the lower lip turning downward and is broader. Flowers are pale yellow with purple spots and have protruding stamens. Below the cluster(s) of flowers are conspicuous pale lilac or cream-colored bracts that make the entire flowering head quite attractive.

Common in dry, open, sandy ground, including fields and roadsides, sometimes becoming weed-like.

PINESAP

HORSE-MINT

SICKLE-LEAVED GOLDEN ASTER

Chrysopsis falcata Aster or Composite family

Early July to mid-September
4"- 16" high

A small, low, fibrous rooted, perennial, aster-like, golden yellow, flowering herb. Stems branching and rather white-woolly. The small, narrow, stiff, gray-green, parallel-veined leaves are crowded together and are distinctive in being somewhat curved, or "sickle-shaped." The rich, golden yellow flowers are about 1" across, so they appear quite large for the size of the plant.

Local in dry sands and dry, sandy roadsides in the pine barrens. Distribution limited to sandy areas near the coast from southeastern Massachusetts, including Cape Cod and Nantucket Island, to the pine barrens of New Jersey.

CRANE-FLY ORCHID

Tipularia discolor Orchid family

Late June through August
8"- 20" high

From a series of short, vertical, thickened, underground stems (corms), this perennial herb first sends up a leafless flowering scape bearing a loose spike of many inconspicuous, greenish, yellowish, or purplish, often mottled, purple-veined flowers. All parts of the flower are narrow and relatively long and the three-lobed lip bears, at its base, a hollow tube, or spur, nearly 1" long. To quote from Stone, 1910, "The absence of any foliage and the spidery character and obscure coloring of the flowers make it an exceedingly difficult plant to detect." Later, after the flowers have withered away, a single, egg-shaped, green (purplish underneath) leaf appears during the fall, persists over winter, and withers in the spring before the new flowers appear. The new flowering scape grows from the same corm that produced the leaf. This corm then develops another corm, to one side, which becomes the source of the next leaf and scape.

Rare. In rich wooded soils. Not regarded as a pine barrens species but is known from at least one pine barrens location.

SICKLE-LEAVED GOLDEN ASTER

CRANE-FLY ORCHID

BUTTONBUSH

Cephalanthus occidentalis Madder family

Early or mid-July to early or mid-August
3'- 18' or 20' high
Commonly 3'- 5' high

A native upright, or obliquely upright, widely branching shrub, rarely tree-like, with smooth, brown-gray bark. Leaves opposite or in whorls of three. Leaves egg-shaped or elliptical, sharp-pointed, rounded or narrowed at base, without marginal teeth or finely so, deep green above, strongly veined, paler beneath, and smooth. A conspicuous bush when in full bloom, with small, dull white flowers in dense, ball-like (spherical) heads, about 1" in diameter, that are terminal at the ends of both the main stems and the many branching stems. Individual flowers are tubular-funnel-form, with four erect or spreading lobes, and with four stamens. Styles very slender, about twice the length of the petals, numerous, and appear pin-like on the flower ball. Very fragrant.

Frequent in swamps, low ground, and shallow waters along the edges of ponds and streams.

ST. ANDREW'S CROSS

Hypericum hypericoides = Ascyrum hypericoides
St. John's-wort family

Early July to early September
4"- 10" high

A very low, smooth, sprawling, diffusely branching, almost sub-shrubby herb with opposite, small, deep green leaves that are oblong or narrowly ovate, stemless, and thin. Flowers are distinguished by having four sepals and four petals, with the sepals in two sizes: two large, broad, outer ones enclosing two very small, narrow, inner ones, which may even be lacking. (Compare with St. Peter's-wort on page 136.) Flowers terminal or at leaf angles, with the four narrow, lemon-yellow petals forming an oblique cross.

Common to locally abundant in dry or moist, sandy soils, including roadsides.

July 113

BUTTONBUSH

ST. ANDREW'S CROSS

NUTTALL'S LOBELIA
Lobelia nuttallii Bluebell or Bellflower family

Early July to early September
8"- 30" high

A small, slender, erect perennial herb with a very slender, smooth, single thread-like stem, occasionally with a few erect branches. Lower leaves narrowly oblong, lance-shaped, upper leaves smaller and linear. Leaves nearly without teeth. Flowers in very open, loose spike toward top of stem. Flowers small, light blue, with white center and two greenish spots. Flowers two-lipped, the upper lip with two lobes, the lower lip with three lobes.

Frequent to common in moist, sandy ground throughout the pine barrens.

WHITE FRINGED ORCHID
Habenaria blephariglottis = *Platanthera blephariglottis*
Orchid family

Mid-July to mid-August
1'- 3' high

From a thick cluster of roots, the smooth, slender stem of this perennial herb arises to bear one to three linear or lance-shaped lower leaves and several much reduced upper ones, plus a compact spike of 5-15 pure- or creamy-white flowers. Lower leaves may be 4"- 8" long. The two most conspicuous parts of the flower are the lower lip and the spur. The narrow, lower lip, actually a petal, may be 1/2" or more long and is finely fringed along its margins, but the lip is not divided and so remains entire. The slender spur that drops down behind the lip is 1"- 1 1/4" long.

Frequent in pine barren bogs, swamps, and low, damp, grassy meadows and savannas. Occasional and local in similar roadside habitats.

YELLOW FRINGED ORCHID
Habenaria ciliaris = *Platanthera ciliaris*
Orchid family

Late July to late August
1'- 2 1/2' high

Another beautiful species that is virtually identical to the above except for its golden or orange-yellow flowers. Formerly known from a few isolated locations. Now exceedingly rare. This author knows of only one very small population within the pine barrens (not illustrated).

NUTTALL'S LOBELIA

WHITE FRINGED ORCHID

PICKERING'S MORNING-GLORY
Stylisma pickeringii = Breweria pickeringii
  Morning-glory family

Mid-July to late August
Not over 2"- 3" high
Vines up to 3'- 6' long

A very low, trailing, vine-like, herbaceous perennial with slender, branching stems that lay on the sandy soil and produce very narrow, linear, almost needle-like leaves, 1" or 2" in length. Flowering stalks develop out of the axils of the leaves and may be as long as, or longer than, the leaves, each stalk bearing two or more narrow, linear, leaf-like bracts and one to five, more often one to three, funnel-shaped, white, morning-glory type flowers, 1/2" long by 1/2"- 3/4" across.

Rare. Confined to dry, white, sandy stretches in the pine barrens and known from only a few stations.

PICKERING'S MORNING-GLORY

GOLDENRODS
Solidago spp. Aster or Composite family

Mid-July (early species) to early October (later species)
1' or 2'- 4' or 5' high

Aster or Composite family members, including asters, goldenrods, bonesets or thoroughworts, and others, dominate the parade of late summer and fall wild flowers in the pine barrens. These are known as composite flowers because what appear to be single flowers are actually composed of many very small flowers crowded together into single flowering heads.

Goldenrods are perennial herbs that develop each year from thick, creeping, underground stems or thick crowns. This is a large, diverse group, with all members having compound (composite) flower heads composed of tiny, tubular flowers (florets) on a central disk, surrounded by a few to several, seldom more than 10-12, ray flowers which, in most species, are a shade of yellow. An exception is silver-rod, *Solidago bicolor*.

Identification of many species is difficult, so to help field identification of the 10 pine barren species that follow, these are divided into three groups according to their flowering patterns. In addition, careful attention must be given to their leaves, specifically whether these are basal or stem leaves. Basal leaves usually are large, differ in size and shape, in their margins (toothed or entire), in their veins, in the length and width of their stalks (petioles), and how they are attached to the main stem. Stem leaves are usually much smaller and become progressively reduced up the stem, but here also attention needs to be paid to their size, their margins, their veining, how crowded or sparse they are on the stem, and their method of attachment to the stem.

Here are the three different flowering patterns or arrangements:

Group I. **Wand-like** or club-like. Flower heads in long, slender, cylindrical spikes at the tops of stems, below which, usually, are small clusters of flowers in axils of upper stem leaves. Never in one-sided clusters. Species in this group are *Solidago erecta, S. stricta, S. bicolor,* and *S puberula*.

Group II. **Plume-like** or elm-branched. Flower heads located only along one (upper) side of long, branching, often curved, spreading, plume-like branches. In some, flower heads are at ends of shortened branches, all branches together forming a terminal plume. These species are *Solidago odora, S. nemoralis, S. rugosa, S. fistulosa,* and *S. uliginosa*.

Group III. **Flat-topped** or round-topped. Flower heads crowded at or near tips of erect branches, the whole forming a flat-topped or dome-shaped arrangement. There is only one pine barrens species, *Solidago tenuifolia*, but a similar *S. graminifolia* has become introduced.

FRAGRANT or SWEET GOLDENROD
Solidago odora Aster or Composite family
(A group II, plume-like, goldenrod)

Mid-July to late August
2'- 4' or 5' high

Stem slender, smooth, often reclining. Leaves chiefly on middle to upper stem, upper leaves very small. Leaves narrow, lance-shaped, entire (without teeth), indistinctly three-ribbed (veined), closely attached to stem, without stalks of their own. Leaves shining, light green, dotted with minute glands (use hand lens). Stems and leaves anise-scented when crushed. Flowering head in a plume-like arrangement. Flower heads small, in small, curved, one-sided branches, with three to five golden-yellow rays.

Frequent, generally distributed in dry, open sands and woods. First pinelands goldenrod to bloom.

SLENDER or ERECT GOLDENROD
Solidago erecta Aster or Composite family
(A group I, wand-like, goldenrod)

Early August to late September
1'- 4' high

Stem and leaves usually smooth. Basal leaves broadly lance-shaped, but broadest beyond middle and tapering downward into stalks that attach to main stem. Upper stem leaves narrow, lance-shaped, and spreading or ascending. Flowering head in wand- or club-like arrangement. Terminal flower heads long, narrow, and cylindrical, not at all one-sided. Flower heads sometimes also on long, straight, or arching cylindrical branches. Flowers with five or six to nine light yellow rays. Bracts surrounding individual flowers blunt.

Occasional, but generally distributed, in dry, open woods and thickets.

July 119

FRAGRANT or SWEET GOLDENROD

SLENDER or ERECT GOLDENROD

WAND-LIKE GOLDENROD

Solidago stricta Aster or Composite family
  (A group I, wand-like, goldenrod)

Mid-August to late September
1'- 5' high

Stem slender and smooth. Basal leaves firm, thick, lance-shaped, broadest beyond middle, tapering to their base, and entire (not toothed). Stem leaves greatly reduced, entire, hug stem, with upper ones numerous, erect, often not much more than bracts. Flowering head in wand- or club-like arrangement. Flowering clusters very long, narrow, terminal, sometimes nodding at tip, and usually unbranched, but occasionally also on short, curved, one-sided branches. Individual flowers with three or five to seven golden rays.

Somewhat restricted to damp, sandy pinelands. Not known in New Jersey outside of pine barrens areas.

FIELD or GRAY GOLDENROD

Solidago nemoralis Aster or Composite family
  (A group II, plume-like, goldenrod)

Mid-August to late September
1'- 4' high

Stem simple, unbranched, and finely but densely hairy, with downy, grayish pubescence. Leaves rough, thick, grayish-green, and indistinctly three-ribbed. Basal leaves broader, indistinctly three-ribbed, toothed toward tips, gradually tapering down to their stalks. Upper leaves narrow and pointed, with tiny leaflets in axils of larger leaves. Flowering head in a plume-like arrangement. Flowers clustered in narrow flowering heads and distinctly one-sided. Individual flowers with 5-9 deep, golden-yellow rays.

Frequent in dry, sterile fields and open woods, often becoming weed-like in old fields.

WAND-LIKE GOLDENROD

FIELD or GRAY GOLDENROD

WRINKLE-LEAVED or ROUGH-STEMMED GOLDENROD

Mid- or late August to late September
1'- 5' high

Solidago rugosa Aster or Composite family
(A group II, plume-like, goldenrod)

Plant often branched, like an elm, at top. Stem straight, cylindrical, densely hairy, and thickly set with numerous, crowded leaves. Lower leaves rough, broad, jagged-toothed, and taper to long, margined stalks. Stem leaves wrinkled, veined, deeply and sharply toothed, and hairy underneath on midrib and veins. Upper stem leaves smaller. Flowering head in a plume-like arrangement. Flower heads light golden-yellow, on curved, one-sided branches. Individual flowers with 6-11 rays (not illustrated).

Generally distributed but only occasional in dry fields and thickets in the pine barrens.

PINE-BARREN GOLDENROD

Mid- or late August to late September
2'- 5' high

Solidago fistulosa Aster or Composite family
(A group II, plume-like, goldenrod)

Stem and leaves hairy. Leaves chiefly on stem, numerous, crowded as they develop on stem. Leaves broad based, somewhat clasping, and strongly hairy on midrib underneath. Leaves elliptically-shaped, feather-veined, entire or sparsely toothed. Flowering head in a plume-like arrangement. Flowers on curved, one-sided branches with 7-12 rays (not illustrated).

Frequent and generally distributed in moist, sandy soils and swamps.

SWAMP or BOG GOLDENROD

Mid- or late August to early October
2'- 5' high

Solidago uliginosa Aster or Composite family
(A group II, plume-like, goldenrod)

Stem stout, smooth. Leaves thick, smooth, lance-shaped, may or may not be shallowly toothed. Basal leaves very long, up to 12", tapering to a winged or margined stalk that embraces stem. Stem leaves progressively reduced. Flowering heads in a plume-like arrangement. Flowers on short stalks, crowded on slender stems or branches that may be either curved and one-sided or straight and not one-sided. Individual flowers with 1-8, more often 5-6, small, light golden rays (not illustrated).

Generally distributed, but infrequent, in pineland bogs and swamps.

SILVER-ROD
Solidago bicolor Aster or Composite family
(A group I, wand-like, goldenrod)

Mid- or late August to early October
1'- 3' high

Stem upright. single or branched, grayish-hairy. Basal leaves broadly lance-shaped or elliptic, rough-hairy, toothed or entire, feather-veined with ribs hairy underneath. Upper leaves entire. Flowering head in wand- or club-like arrangement. Flowers in long, narrow, cylindric, terminal clusters, not at all one-sided. Flower heads small. Central tubular florets creamy-yellow, outer 3-12, more often 7-9 ray flowers whitish. Our only "white" goldenrod.

Frequent in dry, open, sandy woods and roadsides.

SILVER-ROD

DOWNY GOLDENROD

Early September to early October

Solidago puberula Aster or Composite family   1'- 4' high

(A group I, wand-like, goldenrod)

Stem and leaves covered with fine, minute hairs. Stem often purplish. Lower leaves lance-shaped or oblong, toothed, and stalked, with larger ones broadly lance-shaped to elliptic, broadest beyond the middle, and tapering to the base. Stem crowded with 12-60 narrow, linear, entire (untoothed) leaves. Flowering head in wand- or club-like arrangement. Flower cluster long and slender, not at all one-sided. Flowers with 9-16 large, bright yellow rays. Bracts surrounding flower heads acute or pointed (not illustrated).

Frequent and generally distributed in dry, sterile, peaty sands that often have been disturbed. Last goldenrod to come into bloom in the pine barrens.

SLENDER-LEAVED or
FLAT-TOPPED GOLDENROD

Mid- or late August to early October

1'- 3' high

Euthamia tenuifolia = *Solidago tenuifolia*
    Aster or Composite family

Stems slender, smooth. Leaves small, very narrow, linear, almost grass-like, usually only one-ribbed (veined), but may have three indistinct veins. Leaves fragrant and covered with minute resin dots (use hand lens). Tufts of smaller leaves in axils of larger leaves. Flowering heads at ends of numerous erect branches in a flat- or round-topped or dome-shaped arrangement. Flowers numerous, small, with only 3-7 disk flowers in a cluster, each yellow flower with 6-12 minute rays.

Frequent in dry or damp sandy soils.

GRASS-LEAVED or
FLAT-TOPPED GOLDENROD

Mid- or late August to early October

1'- 5' high

Euthamia graminifolia = *Solidago graminifolia*
    Aster or Composite family

This is a similar flat-topped species that has made its way into the pine barrens. This may be distinguished by having larger leaves with three to five, sometimes up to seven, parallel veins (not illustrated).

July 125

SLENDER-LEAVED or
FLAT-TOPPED GOLDENROD

COPPERY ST. JOHN'S-WORT
Hypericum denticulatum St. John's-wort family

Mid-July to early September
6"- 30" high

Smooth, slender, perennial herb. Stem distinctly four-angled or four-ridged. Leaves small, narrow, pointed or elliptic, and may have small teeth along their edges, thus "denticulate." Flowers small, numerous, on leafless branches, with five coppery-yellow petals. Flowers showy, between 1/3" and 1/2" across.

Frequent in pine barren bogs and swamps.

SHORT-LEAVED MILKWORT
Polygala brevifolia Milkwort family

Mid-July to mid-October
2"- 3" up to 12" high

This small, low, erect annual herb has slender stems and branches, the latter sometimes taller than the main stem. Leaves are small, narrow, linear to lance-shaped, usually in sparingly distributed whorls of three to five leaves. On top of the main stem and each (usually) of its branches is a small, dense, clover-like head of tightly clustered, pale rose-purple flowers. The distance between the base of the flower head and the topmost whorl of leaves may be as much as, or more than, one inch.

Common in moist, sandy ground such as open cedar and sphagnum bogs and moist, grassy roadsides throughout the pine barrens.

CROSS-LEAVED MILKWORT
Polygala cruciata Milkwort family

Mid-July to mid-October
3" or 4" up to 18" high

A similar, erect, annual herb but with a somewhat more stout stem that may be squarish or almost winged at the angles. Leaves narrow, in clusters of four (usually), thus forming crosses. Flower heads blunt, on very short stems, the topmost whorl of leaves almost directly underneath flower head, with almost no space between the two (compare with short-leaved milkwort above). Flowers purplish to greenish-white (not illustrated).

Common along margins of swamps and other damp, sandy ground in the pine barrens.

COPPERY ST. JOHN'S-WORT

SHORT-LEAVED MILKWORT

RATTLESNAKE-PLANTAIN
Goodyera pubescens Orchid family

Mid- or late July to early August
6"- 18" or 20" high

This perennial herb develops from a small, branching, fleshy rootstalk and produces a single (usually), stout, densely woolly, flowering scape that bears several small, lance-shaped scales or bracts. At the base of this scape is a rosette of 1"- 3", pointed, egg-shaped, dark green or blue-green leaves, conspicuously veined with white, the middle vein broadly so. Flowers white or greenish-white, in a dense, terminal spike on which the individual flowers grow on all sides of the stem and thus are arranged in sort of a cylinder on the spike. Upper sepals and side petals of flowers are united to form an ovate hood. Lower lip is short and blunt.

Occasional in dry woods, usually where there is a good accumulation of organic matter on the forest floor. Not considered to be a pine barrens species, yet is known from several pine barren locations.

GREEN WOOD ORCHID
Habenaria clavellata = *Platanthera clavellata*
Orchid family

Mid- or late July to mid-August
6"- 18" high

The smooth stem of this small perennial herb develops from a cluster of thick roots and bears a single, rarely two, narrow, tapering leaf on its lower third, plus 1-3 tiny, scale-like bracts above the leaf. The flowering spike is short and broad and is composed of a loose, open cluster of 5-12 small, rather insignificant greenish- or yellowish-white flowers, each with tiny sepals and petals, a short, blunt, wedge-shaped lower lip, and a long, slender spur. This spur, characteristically, is curved upward and around to one side so that the spur is almost sideways and the flower seems distorted. The spur may also have a dilated, or swollen, tip.

Common in low, wet, acid woods, bogs, and swamps. Often found in dark cedar and sphagnum woods where, because of the shade and because the flowers are so small and inconspicuous, they are often overlooked.

July 129

RATTLESNAKE-PLANTAIN

GREEN WOOD ORCHID

SWEET PEPPER-BUSH
Clethra alnifolia White alder family

Mid- or late July to early September
3'- 10' high

This much branched, woody shrub has ascending, slender stems with dull, dark brown bark. Twigs covered with a minute, gray hairiness when young. The deep green, alternate, 1"- 3" leaves, paler underneath, are strongly ovate, pointed at the apex, narrowed or tapering at the base, sharply toothed from mid-leaf to tip, smooth or nearly so, and on very short stems. Flowers small, spicy fragrant, white or pale pink, with five petals, the central pistil protruding beyond the stamens, in slender but dense terminal clusters 3"- 5" long. Seed pods small and almost spherical.

Common to locally abundant in low, moist, sandy swamps, woods, and around the margins of ponds, reservoirs, and streams throughout the pine barrens. One of the very last shrubs to come into bloom, and when in full bloom, its strong, pleasant, spicy fragrance literally fills the atmosphere.

SCLEROLEPIS
Sclerolepis uniflora Aster or Composite family

Mid- or late July to early September
4"- 12" above water surface

This small, slender aquatic perennial herb develops from a creeping, underground stem that usually is in sandy, underwater soil. Its mostly unbranched, usually underwater stem(s) bears small, narrow, entire leaves in whorls of from 4-6. Its erect flowering stalks are above water, and each is terminated by a single, small, button- or disk-like head of pale pink flowers. Individual flowers are all tubular, without any rays. The small, round, pink heads that poke above the water surface have been likened to the appearance of English daisies.

Occasional in shallow waters along the margins of streams and, rarely, in remote wet bogs and savannas.

ASTERS
Aster spp. Aster or Composite family

Mid-July to late October

Asters, like goldenrods, are common fall, composite-type, pine barrens wild flowers. All have compound heads composed of small, yellow, tubular flowers (florets) in a central disk, surrounded by outer ray flowers. In some species, the central disk florets turn dull magenta or reddish with age. The outer ray flowers are bluish, lilac, purple, pink, or even whitish.

SWEET PEPPER-BUSH

SCLEROLEPIS

SHOWY ASTER
Aster spectabilis Aster or Composite family

Mid- or late July to late September
1'- 2' high

Stems stiff, erect or rise obliquely (ascending), simple or branched above, from stout perennial roots. Spreads by slender runners. Stems slightly rough near base. Leaves thickish, firm. Basal and lower leaves oblong-lance-shaped, 3"- 5" long, sparingly toothed, narrowed at base into long, slender stalks. Upper leaves entire, linear-oblong, pointed, attached directly to stem. Flower heads few to numerous, up to 1 1/2" across, very showy, with 15-30 long, 3/4" or more, bright violet or deep blue-violet rays. Tips of bracts surrounding flower heads spreading.

Common in dry, sandy pine barrens.

SLENDER ASTER
Aster gracilis Aster or Composite family

Mid- or late July to early September
6"- 24" high

Small, stiff herb. Several stems may arise from an enlarged rootstalk. Leaves firm, thick, rough, elliptical, entire (not toothed), basal leaves stalked. Stem leaves narrower and mostly attached directly to stem, without stalks. Flower heads few to several, each with 8-14 short, less than 1/2", blue-violet to rose-purple rays that often may be rather pale (not illustrated).

Common in dry, sandy woods of pine barrens.

BOG ASTER
Aster nemoralis Aster or Composite family

Mid- August to late September
6"-24" high

The slender stems of this bog herb may be covered with downy, slightly sticky hairs. Plant spreads by creeping, underground, thread-like runners. All leaves are on stems, none are basal. Leaves numerous, narrow, lance-shaped, taper at both ends, attach directly to main stem, and have rough but untoothed margins that roll slightly inward. Flower heads large, up to 1-1/2" across, either solitary or several on slender stalks. Rays light violet-purple, up to 1/2", and numerous, from 13-27 in number. Floral bracts narrow, sharply pointed, and may be purple tinged.

Frequent, and considered to be *the* aster of cedar bogs and swamps.

SHOWY ASTER

BOG ASTER

LATE PURPLE ASTER
Aster patens Aster or Composite family

Mid- August to early October
1'- 3' high

Stems slender, stiff, somewhat rough-hairy, with several spreading branches toward summit. Leaves ovate-oblong, without stalks, conspicuously clasping main stem with basal lobes that almost encircle stem. Leaves rough-hairy above, thick, rather rigid, with rough but entire margins. Flowers with 20-30 deep blue-violet to light violet-purple rays nearly 1/2" long. Bracts surrounding flower heads pointed and spreading, with green tips. One authority indicates the New Jersey form is *A. p. gracilis*, with plants on average smaller and more slender than in the nominate form.

Common in dry, sandy soils, including roadsides.

SILVERY ASTER
Aster concolor Aster or Composite family

Mid- or late August to early October
1'- 2' high

Stem slender, wand-like, thinly silky-hairy. Leaves oblong, silky-hairy on both sides. Leaves numerous, directed obliquely upward, without stalks, only slightly clasping stem. Flowers in a narrow, elongated pattern, sometimes with a few short branches. Flowers small, up to 3/4", with 8-16 lilac to bluish-lilac rays.

Infrequent to rare in dry, sandy, barren soils.

BUSHY ASTER
Aster dumosus Aster or Composite family

Mid- or late August to early October
1'- 3' high

Stems that arise from creeping, underground stems are usually smooth, or slightly fine-hairy, and sometimes brownish. Stems usually multi-branched and bushy. Leaves on main stem narrow, linear to lance-shaped, bluntly pointed, and attached directly to stem. Leaves on flowering branches much reduced to hardly more than bracts, and numerous. Flowers small, 1/2"- 3/4", with 15-25 pale lavender, bluish-lavender, or almost whitish rays, on long stalks at the ends of branches. Bracts of flowering heads narrow, not spreading.

Frequent in dry (usually) or moist, sandy soils.

LATE PURPLE ASTER

SILVERY ASTER

BUSHY ASTER

BRISTLE-LEAVED or STIFF-LEAVED ASTER

Early September to mid- or late October
6"- 24" high

Aster lineariifolius Aster or Composite family

Stems stiff, wiry, often growing in tussocks, from a stout, vertical, underground stem. Stems very finely downy, with numerous stiff, rough-edged, narrow, needle-like, one-nerved leaves. Flowers rather large, showy, either solitary (frequent), or in small clusters at tops of stems, with 10-20 light violet or blue-violet, rarely violet-white, rays.

Common in dry, sandy soils and open woods.

NEW YORK ASTER

Early September to late October
1'- 4' high

Aster novi-belgii Aster or Composite family

A large, stout (usually) herb that develops from long, creeping, underground stems. Stems smooth or minutely hairy. Leaves chiefly on stems, smaller, upper ones attached directly to stem, slightly but not deeply clasping. Stem leaves long, up to 3", smooth, narrowly lance-shaped to linear, with or without teeth, and lower leaves slightly toothed. Flowers large, up to 1/4", showy, with 15-24 pale to deep blue-violet to deep violet rays. Flower bracts with outward curving tips. Highly variable (not illustrated).

Frequent in moist soils such as damp fields, meadows, thickets.

ST. PETER'S-WORT

Mid- or late July to mid-September
1'- 2' high

Hypericum stans = *Ascyrum stans*
St. John's-wort family

An erect or semi-erect, usually single-stemmed, almost shrubby herb with thickish, opposite, paired, oval-oblong-elliptic leaves that partially clasp stem. Stems somewhat two-edged, with two wing-like ridges. Leaves often have an indented base. Flowers distinguished by having four sepals of uneven sizes: two large, broad, outer ones enclosing two smaller, narrow, inner ones (compare with St. Andrew's cross on page 112). Flowers at the tips of the stems and in the axils of the leaves, with four large lemon- or almost orange-yellow petals that may be 1/2" or more long.

Occasional to common in dry or damp, sandy soils.

July 137

BRISTLE-LEAVED or
STIFF-LEAVED ASTER

ST. PETER'S-WORT

CRESTED YELLOW ORCHID
Habenaria cristata = Platanthera cristata
   Orchid family

Late July to late August
1'- 2 1/2' high

As with the white and yellow fringed orchids, this perennial herb develops from a dense cluster of fibrous roots that sends up a smooth, slender stem that bears one to three narrow, lance-shaped lower leaves, diminishing to the size of bracts up the stem. Atop this stem is a small, dense cluster of small, orange-yellow flowers, the cluster, at least in pine barrens species, considerably shorter than the clusters of either the white fringed or the southern yellow orchids. Individual flowers have narrow, fringed ("bearded" or "crested") lateral petals, and a deeply fringed lower lip. The spur is short, with both the lip and the spur averaging only about 1/4" long.

Very local in bogs and along the edges of open, flood plain savannas.

SOUTHERN YELLOW ORCHID
Habenaria integra = Platanthera integra
   Orchid family

Very early August into September
1'- 2' high

As with the white and yellow fringed and crested yellow orchids, this perennial herb develops from a thick cluster of fibrous roots that sends up a smooth, slender stem that bears one or two narrow, lance-shaped leaves below, the upper one smaller, and several erect, but reduced, bract-like leaves up the stem. The flowering head of golden yellow flowers is almost cylindrical, and usually is considerably taller than the crested yellow orchid, with the two species frequently found together in the same habitat. Individual flowers very small but numerous and densely clustered. Lip is tongue-shaped and toothed on the edge, but is *not* fringed, distinguishing this species as the only fringeless rein orchid in the pine barrens. Spur short, approximately 1/4", tapering toward the tip.

Very local in bogs and in open flood plain savannas.

CRESTED YELLOW ORCHID

SOUTHERN YELLOW ORCHID

GRASS-LEAVED BLAZING STAR
Liatris graminifolia Aster or Composite family

Early August to late September
1'- 3' high

Slender, unbranched spikes of this herb rise from perennial, underground rootstalks and bear both basal and stem leaves. Basal leaves are narrow-bladed and stalked, their flat stalks edged with hairs, but these leaves may be lacking by flowering time. Stem leaves are mostly narrow, singly attached to stem, grass-like, and may be in close spirals. Leaves may be marked with resinous dots (use hand lens). Flowering spikes consist of from 5-15 rose-purple to lavender-lilac flowers, loosely arranged in open spike- or wand-like clusters. Individual flowers small, tubular, five-lobed, in small, cylindrical but narrow-based heads. Stems and flower stalks hairy. Leafy bracts below flower heads narrowly oblong, blunt, edged with hairs, and either ridged or striped on undersides.

Frequent in dry to moist, sandy pinelands.

NARROW-LEAVED SUNFLOWER
Helianthus angustifolius Aster or Composite family

Early August to late September
2'- 5' high

A perennial herb from slender rootstalks. Stem slender, branched above, usually somewhat hairy below. Leaves firm or stiff to rigid, slightly rough, linear, without teeth, and attached directly to stem. Leaves long and very narrow, from 2"- 7" long, but only 1/6"- 1/3" wide, margins often becoming rolled over with age or in drying. Lower leaves opposite, upper ones alternate. Flower heads large with 12-20 yellow ray flowers and a central disk of dark, purplish, tubular flowers (florets). Entire head 2"- 3" across.

Occasional to frequent in bogs, swamps, and damp thickets.

GRASS-LEAVED BLAZING STAR

NARROW-LEAVED SUNFLOWER

MARSH ST. JOHN'S-WORT
Triadenum virginicum = Hypericum virginicum
St. John's-wort family

Early August to early September
1'- 2' high

An erect, perennial herb that develops from an underground stem. Leaves paired, close-set, without stalks, light green with sepia dots (use hand lens), and with a slight bloom underneath. Leaves are broad, embrace the stem, are 2-3 times longer than wide and are blunt at tips. Flowers, in small terminal clusters, as well as in upper leaf axils, are pinkish or flesh-colored, with three groups of three golden-yellow stamens (total nine), separated by three large orange glands. This is our only pink St. John's wort. An interesting feature of these flowers is that they do not open until late in the afternoon.

Common in open bogs, marshes, and swamps, as well as along the grassy edges of ponds and reservoirs.

SWAMP LOOSESTRIFE or WATER-WILLOW
Decodon verticillatus Loosestrife family

Early August to early September
2'- 8' or 10' high

A somewhat shrubby, semi-aquatic perennial with nearly smooth, reclining, wand-like, angular, four- to six- sided stems that become, near their base, almost woody and are often covered with a leathery or corky growth. Stems grow so tall and are so supple that their upper ends frequently bend over and down so low that their tips reach the mud and take root. Lance-shaped or willow-leaf-shaped leaves are paired on the lower stem but are in whorls of three (mostly) or four on the upper stem, and have short stalks. Numerous bell-shaped, purplish-pink or magenta flowers are densely clustered in the upper leaf axils. Individual flowers have five wedge or lance-shaped petals with projecting stamens.

Common in swamps and along the margins of ponds and slow-moving streams.

August 143

MARSH ST. JOHN'S-WORT

SWAMP LOOSESTRIFE or
WATER-WILLOW

MARYLAND GOLDEN ASTER
Chrysopsis marianra Aster or Composite family

Early August to early or mid-September

1'- 2 1/2' high

Stems stout, loosely hairy, nearly erect or rising obliquely (ascending), often more than one stem rising from a single, stout, perennial root. Stems may be many-branched into a flat- or round-topped summit. Stem and leaves silky with soft hairs when young but become smooth with age. Lower leaves oblong or spatulate-shaped, blunt, widest between the middle and the tip, and narrowed down to stalks. Upper leaves gray-green, oblong or lance-shaped, blunt or pointed, and attached directly to stem without leaf stalks. Flower heads often numerous, both central disk flowers and outer ray flowers bright golden yellow, 3/4"- 1" across. Both the stalks bearing the flower heads and the flower head bracts beneath the flowers slightly sticky.

Common to abundant in dry, sandy clearings and open woods and along dry roadsides.

MARYLAND GOLDEN ASTER

BONESETS or THOROUGHWORTS
Eupatorium spp. Aster or Composite family

Late July (*E. pilosum*), others all early August to early or mid- September
1'- 4' high

Like goldenrods and asters, bonesets or thoroughworts are a major component in the parade of late summer and fall composite-type wild flowers in the pine barrens. Although not as showy and colorful as either the goldenrods or asters, all are tall, erect, perennial herbs with numerous branches near the tops of their stems. All pine barrens species have white, or grayish-white, tubular flowers of the central disk type. None have any ray flowers. In most cases, the small, fuzzy, flowering heads are clustered at the tops of branches to form a generally flat or round-topped inflorescence. One species, *E. hyssopifolium* (see page 146), has its leaves in whorls of threes or fours. In other species, leaves are usually opposite, in pairs, and this provides a good way to identify these species. Refer to *A Field Guide to the Pine Barrens of New Jersey*, page 209, for diagnostic illustrations.

ROUGH BONESET
Eupatorium pilosum = *E. rotundifolium*
var. *saundersii* Aster or Composite family

Late July to mid-September
2'- 4' high

Plant rough, stem hairy. Leaves without stalks, thus closely attached to stem. Lower leaves opposite, oblong, somewhat blunt, with few (3-12) coarse teeth on each margin. Upper leaves nearly entire, without teeth, linear or lance-like. About five individual flowers (florets) to a single flowering head (not illustrated).

Frequent in damp or moist to wet sandy soils and bogs. This is the earliest boneset to come into bloom, usually beginning to blossom about two weeks earlier than any other boneset.

HAIRY BONESET
Eupatorium pubescens = *E. rotundifolium*
var. *ovatum* Aster or Composite family

Early or mid-August to early or mid-September
2'- 4' high

Stem hairy. Lower, larger leaves opposite, oblong-pointed or egg-shaped, with 12-25 sharp teeth on each margin. Upper leaves toothed (not illustrated).

Common in moist or dry (more often) fields, open woods, clearings.

PINE-BARREN or RESINOUS BONESET
Early or mid-August to early or mid-September
2'- 3' high
Eupatorium resinosum Aster or Composite family

Stems slender, minutely soft-pubescent. Leaves opposite, finely pubescent, sometimes resinous or slightly sticky underneath, narrowly lance-shaped with toothed edges, and with narrowed bases that clasp stem. Flowers 10 or more to a head, in terminal clusters (not illustrated).

Restricted to wet bogs in the heart of the pine barrens.

WHITE BONESET
Early or mid- August to mid- September
Eupatorium album Aster or Composite family
1'- 3 1/2' high

Stem rough-hairy or downy. Leaves opposite, lance-shaped, narrowed to the base, veined, nearly without stalks, coarsely toothed, light green. About five white flowers in each flower head (not illustrated).

Common in dry, sandy, open pine barrens.

WHITE-BRACTED BONESET
Early or mid-August to mid-September
Eupatorium leucolepis Aster or Composite family
1 1/2'- 3 1/2' high

A slender herb with thick, opposite leaves. Leaves narrowly lance-shaped, nearly without stalks, with few small teeth, and with three obscure veins. There may be tufts of smaller leaves in leaf axils. Bracts supporting flower heads white (not illustrated).

Frequent in open, damp sands, peats, bogs, and swamps.

HYSSOP-LEAVED BONESET
Early or mid-August to mid-September
Eupatorium hyssopifolium Aster or Composite family
1'- 3' high

Easily identified by its whorls, in threes or fours, of very narrow, almost linear, grasslike, entire or sometimes slightly toothed leaves, with clusters of smaller leaves in leaf axils. Some upper leaves may be paired or single. About five white flowers in each flower head.

Frequent to common in dry or moist open, sandy fields, woods, roadsides.

ROUND-LEAVED BONESET
Early or mid-August to mid-September
Eupatorium rotundifolium Aster or Composite family
2'- 4' high

Stem hairy. Leaves opposite. Leaf blades broadly ovate, almost as broad as long, palmately veined, semi-clasping around stem, with large, blunt teeth. Five to seven white flowers in each flower head.

Common in dry, open, sandy soils, woods.

August 147

HYSSOP-LEAVED BONESET

ROUND-LEAVED BONESET

BLUE CURLS
Trichostema dichotomum Mint family

Mid-August to mid-September
6"- 18" high

A small, delicate, but sometimes weedy, annual herb remarkable for the very long, curved, blue-stalked stamens of its flowers. Stems erect, stiff, multi-branched, and slightly sticky. Leaves narrowly oblong or lance-shaped, without teeth (entire), without pedicels, slightly sticky, and somewhat aromatic. Flowers numerous but scattered, either singly or in pairs at tips of branches. Individual flowers violet-blue, with four exceptionally long, violet-blue stamens that curl way out beyond the five-lobed (four above, one larger lobe below) flower cup or corolla.

Common in dry, open, sandy fields, waste areas, roadsides.

FERN-LEAVED FALSE FOXGLOVE
Aureolaria pedicularia = *Gerardia pedicularia*
Figwort family

Mid-August to mid-September
1'- 3 1/2' high

An annual, bushy herb with rather slender, erect or spreading, and much branched leafy stems that are somewhat sticky and hairy. Light green leaves are deeply cut into many finely toothed lobes, thus fern-like, and are stemless or nearly so, but lower leaves may have short stalks. Flowers are produced on relatively long stalks (pedicel) from the axils of smaller, upper, bract-like leaves. Individual showy, bright lemon-yellow flowers are bell- or funnel-shaped with five broad, spreading, rounded lobes 1" or more across, with the upper two lobes joined for a somewhat greater part of their length than are the lower three. The outer surfaces and throat of the flower, the four stamens, and the toothed sepals are all finely hairy.

Frequent in dry, open, oak woodlands and roadsides. Plants may be at least partially parasitic on the roots of oak trees and other vegetation.

BLUE CURLS

FERN-LEAVED FALSE FOXGLOVE

THREAD-LEAVED or BRISTLE-LEAVED GERARDIA

Mid-August to mid- or late September
1'- 2 1/2' high

Agalinis setacea = *Gerardia setacea* Figwort family

A slender, erect, branching, annual herb with smooth, wiry stems. Leaves small, paired, narrowly linear, entire (without teeth), and nearly thread-like. Flowers deep pink to magenta-purple, on pedicels longer than 1/4", or as long as or longer than the length of the flower. (The length of the flower stalk is a key difference between this and the following species. Compare below.) Individual flowers small, 1/2"- 3/4" (compare below), broadly funnel-shaped, with five flaring lobes.

Frequent in dry, open, sandy areas and roadsides.

PINE-BARREN GERARDIA

Mid- to late August to late September
1'- 2 1/2' high

Agalinis purpurea var. *racemulosa* =
  Gerardia racemulosa Figwort family

A similar slender, erect, branching, annual herb with smooth, wiry stems. Leaves small, paired, narrowly linear, entire (without teeth), and nearly thread-like, frequently curling, strongly so when dried. Flowers deep pink to magenta-purple, on very short pedicels, less than 1/8" long or only as long as the sepals, from upper leaf axils. Individual flowers large, 3/4"- 1 1/2", broadly funnel-shaped, with five flaring lobes.

Occasional to frequent in open, damp, boggy areas (compare with preceding species).

THREAD-LEAVED or BRISTLE-LEAVED GERARDIA

PINE-BARREN GERARDIA

RATTLESNAKE-ROOT

Mid- or late August to mid-October

Prenanthes serpentaria Aster or Composite family

1'- 3' high

A tall, branched herb with milky juice and narrow flower heads that hang in a loose cluster from the tips of the stems. Stems smooth, green, and leafy. Leaves variable, somewhat roughish, often deeply cut and lobed. Lower leaves thick, broad, three-sided, and remotely toothed. Upper leaves deeply cut, the uppermost lance-shaped with two small, lateral lobes near base or none at all. The small, dull, cream-colored, hanging, bell-shaped flowers are enclosed in a somewhat bristly, hairy, green envelope that may be a trifle magenta-tinted. Flowers from 5-18, usually about 10, to a head, surrounded by eight bracts.

Occasional in dry, sandy barrens, open woods, and roadsides.

PINE-BARREN or
SLENDER RATTLESNAKE-ROOT

Early September to mid-October

1'- 3' high

Prenanthes autumnalis Aster or Composite family

A slender herb with a smooth stem. Basal and lower stem leaves elongate, finely and deeply cut, with winged stalks. Upper stem leaves lance-shaped, toothed, without stalks, and conspicuously reduced upward. From 8-13, usually 10 or 11, flowering heads are arranged in a row up a narrow spike on a slender, unbranched (usually) stem. Lower flowers of inflorescence in leaf axils, upper ones on leafless stem, often all on one side. There are from 8-12 pinkish flowers to an individual flowering head, surrounded by about eight bracts.

Occasional in dry, sandy soils. Not at all common in pine barrens.

August – September   153

RATTLESNAKE-ROOT

PINE-BARREN or
SLENDER RATTLESNAKE-ROOT

JOINTWEED
Polygonella articulata Buckwheat family

Early September to late October
6"- 20" high

An annual herb with smooth, slender, wiry, erect or somewhat spreading, jointed stems that may be only single or much branched. Leaves tiny, narrow, linear, almost threadlike. Leaves so inconspicuous that the plant seems to consist entirely of slender spikes of tiny flowers supported on wiry stems. Flowers small, numerous, either pink, white, or both, on tall, loose spikes. Flowers may become more pinkish after early autumn frosts. Even though individual flowers are small, they are numerous enough to make an attractive showing.

Common in dry, sandy soils almost everywhere in the pine barrens, especially along sandy dikes of old cranberry bogs.

PINE-BARREN GENTIAN
Gentiana autumnalis Gentian family

Early September to mid-October
8"- 24" high

A very slender herb with a weak stem that may occasionally be branched. Leaves opposite, very narrow, linear, up to 2" long. Flowers solitary, usually, or up to two or three, at the tops of stems or branches. The large, deep, tubular flowers open up to display five widely spreading lobes, with pointed plaits between the lobes. Color of flowers varies from bright ultramarine blue to deep blue, with pale stripes and brownish speckles inside corolla.

Occasional and local in moist, sandy habitats.

JOINTWEED

PINE-BARREN GENTIAN

LITERATURE CITED
and other references

Boyd, H. P. 1991. *A Field Guide to the Pine Barrens of New Jersey*. Plexus Publishing, Inc., Medford, NJ

Boyd, H. P. 1997. *A Pine Barrens Odyssey*. Plexus Publishing, Inc., Medford, NJ

Britton, N. L. and A. Brown. 1913, 1952. *An Illustrated Flora of the Northern United States, Canada, and the British Possessions*. Charles Scribner's Sons, NY

Fernald, M. L. 1950. *Gray's Manual of Botany*. 8th ed. American Book Co., NY

Gleason, H. A. and A. Cronquist. 1991. *Manual of Vascular Plants of northeastern United States and adjacent Canada*. 2nd ed. The New York Botanical Garden, NY

Harshberger, J. W. 1916. Reprint 1970. *The Vegetation of the New Jersey Pine Barrens*. Dover Publications, NY

House, H. D. 1934. *Wild Flowers*. The Macmillan Co., NY

Matthews, F. S. 1927. *Field Book of American Wild Flowers*. G. P. Putnam's Sons, NY

Matthews, F. S. 1915. *Field Book of American Trees and Shrubs*. G. P. Putnam's Sons, NY

McCormick, J. 1970. *The Pine Barrens. A Preliminary Ecological Inventory*. New Jersey State Museum, Trenton, NJ

National Geographic Society. 1924. *The Book of Wild Flowers*. The National Geographic Society, Washington, DC

Newcomb, L. 1977. *Newcomb's Wildflower Guide*. Little, Brown and Co., Boston, MA

New Jersey Pinelands Commission. 1980. *Comprehensive Management Plan*. Pinelands Commission, New Lisbon, NJ

Peterson, R. T. and M. McKenny. 1968. *A Field Guide to Wildflowers of Northeastern and North-central North America*. Houghton Miflin Co., Boston, MA

Rickett, H. W. 1963. *The New Field Book of American Wild Flowers*. G. P. Putnam's Sons, NY

Rickett, H. W. 1966. *Wild Flowers of the United States. Vol. 1. The Northeastern States*. The New York Botanical Garden and McGraw-Hill Book Co., NY

Schnell, D. E. 1976. *Carnivorous Plants of the United States and Canada*. John F. Blair, Publisher, Winston-Salem, NC

Snyder, D. B. 1992. *Special Plants of New Jersey*. Office of Natural Lands Management, Div. of Parks & Forestry, N.J. Dep't. of Environmental Protection, Trenton, NJ

Still, C. C. 1998. *Botany and Healing: Medicinal Plants of New Jersey and the Region*. Rutgers University Press, NJ

Stone, W. 1911. *The Plants of Southern New Jersey*. Originally published as Part II of the *Annual Report of the New Jersey State Museum for 1910*. Reprint 1973. Quarterman Publications, Boston, MA

Index to generally-accepted COMMON NAMES

Arbutus, trailing 18
Arethusa 46
Arrow-head 80
Asphodel, bog 76
  , false or viscid 80
Aster, bog 132
  , bushy 134
  , late purple 134
  , New York 136
  , showy 132
  , silvery 134
  , slender 132
  , stiff-leaved 136
  , white-topped, narrow-leaved 78
                        , toothed 78
Asters 130
Azalea, swamp or clammy 68

Bearberry 26
Bladderwort, fibrous 56
  , horned 56
  , pin-like or slender 54
  , purple 58
  , reversed or reclined 58
  , swollen 56
Bladderworts 54
Blazing star, grass-leaved 140
Blue curls 148
Blue-eyed grass, eastern 36
Blue flag, slender 46
Blueberry, highbush 26
Bog-asphodel 76
Boneset, hairy 145
  , hyssop-leaved 146
  , pine-barren 146
  , rough 145
  , round-leaved 146
  , white 146
  , white-bracted 146
Bonesets 145

Broom-crowberry, Conrad's 16
Butterfly-weed 88
Buttonbush 112

Calico-bush 50
Cassandra 20
Centaury, lance-leaved 106
Chaffseed 62
Checkerberry 90
Colic-root 76
Cow-wheat 54
Cranberry, large or American 74

Dandelion, dwarf 34
Dragon's mouth 46

False foxglove, fern-leaved 148
Fly-poison 60
Frostweed 36

Gentian, pine-barren 154
Gerardia, pine-barren 150
  , thread-leaved or bristle-leaved 150
Goat's rue 66
Gold-crest 96
Golden club 18
Golden aster, Maryland 144
  , sickle-leaved 110
Goldenrod, downy 124
  , field or gray 120
  , flat-topped, slender-leaved 124
                    , grass-leaved 124
  , fragrant or sweet 118
  , pine-barren 122
  , slender or erect 118
  , swamp or bog 122
  , wand-like 120
  , wrinkle-leaved or rough-stemmed 122
Goldenrods 117

Grass-pink   62

Heather, beach or woolly   44
, pine-barren or golden   44
Helleborine   100
Horse-mint   108
Huckleberry, black   32

Indian pipe   70
Ipecac spurge   24

Jointweed   154

Ladies' tresses   98
, grass-leaved   100
, little   98
, nodding   100
, spring   98
Lady's-slipper, pink   30
Lambkill   52
Laurel, mountain   50
, sheep   52
Leatherleaf   20
Lily, turk's-cap   102
Lobelia, Nuttall's   114
Loosestrife, swamp   142
, yellow   72
Lupine, wild   30

Magnolia, swamp   52
Mayflower   18
Meadow-beauty   102
, Maryland   104
Milkweed, blunt-leaved   86
, common   86
, red   88
, swamp   90
Milkweeds   84
Milkwort, cross-leaved   126
, orange   72
, short-leaved   126
Moccasin-flower   30
Morning-glory, Pickering's   116
Mountain-laurel   50
Myrtle, sand   28

Orchid, crane-fly   110
, crested yellow   138
, green wood   128
, ragged fringed or green fringed   82
, southern yellow   138
, white fringed   114
, yellow fringed   114

Partridge-berry   50
Pencil-flower   78
Pepperbush, sweet   130
Pickerel-weed   94
Pinesap   108
Pipewort, early or flattened   38
, seven-angled   40
, ten-angled   38
Pipeworts   38
Pipsissewa   82
Pitcher-plant   48
Pogonia, rose   60
, spreading   96
Pond-lily, yellow   42
Prickly pear   68
Prince's pine   82
Putty-root   43
Pyxie   16

Rattlesnake-plantain   128
Rattlesnake-root   152
, pine-barren or slender   152
Red-root   106

St. Andrew's cross   112
St. John's-wort, coppery   126
, marsh
St. Peter's-wort   136
Sand-myrtle   28
Sandwort, pine-barren   64
Sclerolepis   130
Sheep-laurel   552
Silver-rod   123
Skullcap, hyssop   74
Snake mouth   60
Spatterdock   42

Spurge, ipecac 24
Staggerbush 34
Starflower 32
Star-grass, yellow 40
Sundew, round-leaved 94
 , spatulate-leaved 92
 , thread-leaved 93
Sundews 92
Sundrops, narrow-leaved 170
Sunflower, narrow-leaved 140
Swamp candles 72
Swamp-pink 20
Sweet bay 52
Sweet pepperbush 130

Teaberry 90
Thoroughworts 145
Toadflax, blue or old field 28

Turkey-beard 40
Twayblade, large or lily-leaved 58
 , southern 22

Violet, bird's-foot 22
 , lance-leaved 24
 , primrose-leaved 24

Water-lily, fragrant 66
Water-shield 64
Water-willow 142
White-topped aster, narrow-leaved 78
 , toothed 78
Wintergreen 90
 , spotted or striped 82

Yellow-eyed grass, carolina 104
 , slender 104

Index to SCIENTIFIC NAMES

(Names in synonymy are shown in italic type)

Agalinis purpurea var. racemulosa 150
 setacea 150
Aletris farinosa 76
Amianthium muscaetoxicum 60
Aplectrum hyemale 43
Arctostaphylos uva-ursi 26
Arenaria caroliniana 64
Arethusa bulbosa 46
Asclepias amplexicaulis 86
 incarnata 90
 rubra 88
 spp 84
 tuberosa 88
 syriaca 86
Ascyrum hypericoides 112
 stans 136
Aster concolor 134
 dumosus 134
 gracilis 132
 lineariifolius 136

 nemoralis 132
 novi-belgii 136
 patens 134
 paternus 78
 solidagineus 78
 spectabilis 132
 spp. 130
Aureolaria pedicularia 148

Brasenia schreberi 64
Breweria pickeringii 116

Calopogon *pulchellus* 62
 tuberosus 62
Cephalanthus occidentalis 112
Chamaedaphne calyculata 20
Chimaphila maculata 82
 umbellata 84
Chrysopsis falcata 110
 mariana 144

Cleistes divaricata 96
Clethra alnifolia 130
Corema conradii 16
Cypripedium acaule 30

Decodon verticillatus 142
Drosera filiformis 93
    intermedia 92
    routundifolia 94
    spp. 92

Epigaea repens 18
Epipactis helleborine 100
Eriocaulon aquaticum 38
    compressum 38
    decangulare 38
    septangulare 40
    spp. 38
Eupatorium album 146
    hyssopifolium 146
    leucolepis 146
    ovatum 145
    pubescens 145
    pilosum 145
    resinosum 146
    rotundifolium 146
    saundersii 145
    spp. 145
Euphorbia ipecacuanhae 24
Euthamia graminifolia 124
    tenuifolia 124

Gaultheria procumbens 90
Gaylussacia baccata 32
Gentiana autumnalis 154
Gerardia pedicularia 148
    racemulosa 150
    setacea 150
Goodyera pubescens 128

Habenaria blephariglottis 114
    ciliaris 114
    clavellata 128
    cristata 138
    integra 138
    lacera 81

Helianthemum canadense 36
Helianthus angustifolius 140
Helonias bullata 20
Hudsonia ericoides 44
    tomentosa 44
Hypericum denticulatum 126
    hypericoides 112
    stans 136
    virginicum 142
Hypoxis hirsuta 40

Iris prismatica 46

Kalmia angustifolia 52
    latifolia 50
Krigia virginica 34

Lachnanthes caroliniana 106
    tinctoria 106
Leiophyllum buxifolium 28
Liatris graminifolia 140
Lilium superbum 102
Linaria canadensis 28
Liparis liliifolia 58
Listera australis 22
Lobelia nuttallii 114
Lophiola aurea 96
Lupinus perennis 30
Lyonia mariana 34
Lysimachia terrestris 72

Magnolia virginiana 52
Melampyrum lineare 54
Minuartia caroliniana 64
Mitchella repens 50
Monarda punctata 108
Monotropa hypopithys 108
    uniflora 70

Narthecium americanum 76
Nuphar advena 42
Nuphar variegata 42
Nymphaea odorata 66

Oenothera fruticosa 70
    linearis 70

Opuntia humifusa 68
Orontium aquaticum 18

Platanthera blephariglottis 114
    ciliaris 114
    clavellata 128
    cristata 138
    integra 138
    lacera 82
Pogonia ophioglossoides 60
Polygala brevifolia 126
    cruciata 126
    lutea 72
Polygonella articulata 154
Pontederia cordata 94
Prenanthes autumnalis 152
    serpentaria 152
Pyxidanthera barbulata 16

Rhexia mariana 104
    virginica 102
Rhododendron viscosum 68

Sabatia difformis 106
Sagittaria engelmanniana 80
Sarracenia purpurea 48
Schwalbea americana 62
Sclerolepis uniflora 130
Scutellaria integrifolia 74
Seriocarpus asteroides 78
    linifolius 78
Sisyrinchium atlanticum 36
Solidago bicolor 123
    erecta 118
    fistulosa 122
    graminifolia 124
    nemoralis 120
    odora. 118
    puberula 124

    spp. 117
    rugosa 122
    stricta 120
    tenuifolia 124
    uliginosa 122
Spiranthes *beckii* 98
    cernua 100
    praecox 100
    tuberosa 98
    spp. 98
    vernalis 98
Stylisma pickeringii 116
Stylosanthes biflora 78

Tephrosia virginiana 66
Tipularia discolor 110
Tofieldia racemosa 80
Triadenum virginicum 142
Trichostema dichotomum 148
Trientalis borealis 32

Utricularia cornuta 56
    fibrosa 56
    inflata 56
    purpurea 58
    resupinata 58
    spp. 54
    subulata 54

Vaccinium corymbosum 26
    macrocarpon 74
Viola lanceolata 24
    pedata 22
    primulifolia 24

Xerophyllum asphodeloides 40
Xyris caroliniana 104
    torta 104

More Great Books from Plexus Publishing

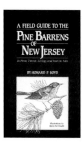

A FIELD GUIDE TO THE PINE BARRENS OF NEW JERSEY
By Howard P. Boyd

"…Howard Boyd has succeeded in the formidable task of bringing together definitive and detailed answers to questions about the Pine Barrens.... a must for anyone who is casually or seriously interested in the New Jersey Pine Barrens." — V. Eugene Vivian, Emeritus Professor of Environmental Studies, Rowan State College

With his 420-page volume, author Howard P. Boyd presents readers with the ultimate handbook to the New Jersey Pine Barrens. Boyd begins his book by explaining and defining what makes this sandy-soiled, wooded habitat so diverse and so unusual.

Each entry gives a detailed, non-technical description of a Pine Barrens plant or animal (for over 700 species), indicating when and where it is most likely to appear. Complementing most listings is an original ink drawing that will greatly aid the reader in the field as they search for and try to identify specific flora and fauna.

1991/423 pp/hardbound/ISBN 0-937548-18-9/$32.95
1991/423 pp/softbound/ISBN 0-937548-19-7/$22.95

A PINE BARRENS ODYSSEY: A NATURALIST'S YEAR IN THE PINE BARRENS OF NEW JERSEY
By Howard P. Boyd

A Pine Barrens Odyssey is a detailed perspective of the seasons in the Pine Barrens of New Jersey. Primarily focused on the chronology of the natural features of the Pine Barrens, this book is meant as a companion to Howard P. Boyd's A Field Guide to the Pine Barrens of New Jersey.

The two books form an appealing collection for anyone interested in the Pine Barrens of New Jersey. The *Field Guide* can be used as a reference tool for the types of flora and fauna and the *Odyssey* as a calendar of what to expect and look for season by season in this beautiful natural area of New Jersey.

1997/275 pp/softbound/ISBN 0-937548-34-0/$19.95

SALT MARSH FLOWERS OF SOUTHERN NEW JERSEY
By Carol E. Newcomb

Salt water, sand flats, and long grasses waving in a hot summer's breeze are only a few of the things we think about when we think of New Jersey's salt marshes and their flowers. In *Salt Marsh Flowers of Southern New Jersey*, the flora of this area is described in great detail by Carol Newcomb, illuminating these gems of the shore. To aid readers in their field searches, drawings of each of the salt marsh flowers discussed are included.

1996/softbound/$8.95

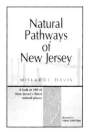

NATURAL PATHWAYS OF NEW JERSEY
By Millard C. Davis

"Laden with keen observations of the unspoiled world, and the feelings these evoke in Davis, *Natural Pathways of New Jersey* represents a rare and genial wedding of science and heart."— *The Central Record*

Natural Pathways of New Jersey describes in eloquent detail over 100 natural places in New Jersey. Davis' descriptions of beaches, forests, and fields include not only the essence of the landscapes, but also portray the animals and vegetation native to the area.

Natural Pathways of New Jersey is divided into sections by county, making it very readable and easy for anyone to find the cited areas. The book includes over 100 original watercolor illustrations by artist Valerie Smith-Pope and two simple trips that anyone can take, showcasing the best that New Jersey has to offer.

1997/271 pp/softbound/ISBN 0-937548-35-9/$19.95

POISONOUS, VENOMOUS, AND ELECTRIC MARINE ORGANISMS OF THE ATLANTIC COAST, GULF OF MEXICO, AND THE CARIBBEAN
By Matthew Landau

Millions of people spend time around the ocean. In doing so most have come into contact with an animal that has a protective mechanism of some sort. Landau explains in detail, with the aid of illustrations, the types of marine animals that exist in the waters of the Atlantic Coast, Gulf of Mexico, and the Caribbean and the particular type of toxins or danger each poses.

1997/218 pp/hardbound/ISBN 0-937548-36-7/$29.95
1997/218 pp/softbound/ISBN 0-937548-33-2/$19.95

PATRIOTS, PIRATES, AND PINEYS
By Robert A. Peterson

"*Patriots, Pirates, and Pineys* is excellent...the type of book that is hard to put down once you open it."—*Daybreak Newsletter*

Southern New Jersey is a region full of rich heritage, and yet it is one of the best-kept historical secrets of our nation. Many famous people have lived in Southern New Jersey, and numerous world-renowned businesses were started in this area as well.

This collection of biographies provides a history of the area through the stories of such famous figures as John Wanamaker, Henry Rowan, Sara Spenser Washington, Elizabeth Haddon, Dr. James Still, and Joseph Campbell. Some were patriots, some pirates, and some Pineys, but all helped make America what it is today.

1998/155 pp/hardbound/ISBN 0-937548-37-5/$29.95
1998/155 pp/softbound/ISBN 0-937548-39-1/$19.95

GATEWAY TO AMERICA
By Gordon Bishop
Photographs by Jerzy Koss

Based on the acclaimed PBS documentary, *Gateway to America* is both a comprehensive guidebook and history. It covers the historic New York/New Jersey triangle that was the window for America's immigration wave in the 19th and 20th centuries. In addition to Ellis Island and the Statue of Liberty, the book fully explores seven other Gateway landmarks including Liberty State Park, Governors Island, the World Trade Center, Battery City Park, South Street Seaport, Newport, and the Gateway National Recreational Area. A must for history buffs and visitors to the Gateway area alike.

1998/152 pp/softbound/ISBN 1-887714-27-8/$19.95
(Originally published by Summerhouse Press—now available exclusively from Plexus)

OLD AND HISTORIC CHURCHES OF NEW JERSEY, VOLUME 2
By Ellis L. Derry

This inspirational book allows us to travel back in time to the days when this country was new—a vast and dangerous wilderness with few roads or bridges, schools or churches. It tells the stories of how our forefathers established their religious communities and houses of worship, often with great hardship and sacrifice.

To be included in this moving history, a church had to be built by the time of the Civil War and still be in use as a church today. This means the book, in addition to its historical appeal, can serve as a guide to readers who would like to pay a visit to many of New Jersey's most interesting churches.

1994/372 pp/hardbound/ISBN 0-937548-25-1/$29.95
1994/372 pp/softbound/ISBN 0-937548-26-X/$19.95

DOWN BARNEGAT BAY: A NOR'EASTER MIDNIGHT READER
By Robert Jahn

"*Down Barnegat Bay* evokes the area's romance and mystery."
—The New York Times

Down Barnegat Bay is an illustrated maritime history of the Jersey shore's Age of Sail. Originally published in 1980, this fully-revised Ocean County Sesquicentennial Edition features more than 177 sepia illustrations, including 75 new images and nine maps. Jahn's engaging tribute to the region brims with first-person accounts of the people, events, and places that have come together to shape Barnegat Bay's unique place in American history.

2000/ 248pp/hardbound/ISBN 0-937548-42-1/$39.95

To order or for a catalog: 609-654-6500, Fax Order Service: 609-654-4309
Web Site: www.plexuspublishing.com

Plexus Publishing, Inc.

143 Old Marlton Pike • Medford • NJ 08055
E-mail: info@plexuspublishing.com